Scientific riddles

Hans smart ... Cultured horse !!

What is the intelligence of animals? Can she think and understand like a person? In the Qur'an, we read about the story of the Prophet Allah Suleiman (as) with Hudhud and with ants, which confirms that animals know and understand what is happening around them.

But what do you think, dear reader, about a horse that can solve math problems and read words !! Can this happen in nonfiction? Read this true story about Clever Hans, who puzzled scientists at one time, and leave his case to you.

At the end of the 19th century, the German mathematician Wilhelm von Osten became interested in the study of intelligence and mental ability, especially in animals, and this study brought him wide popularity. von Osten firmly believed that humanity had grossly underestimated the thinking ability and intelligence of animals, and in order to prove his aforementioned hypothesis, he decided to conduct experiments in teaching animals mathematics !! He began his experiments with three animals: a cat, an Arabian horse, and a bear.

But an Arab horse named Hans showed some promising ability, and with a lot of exercise, the horse taught Hans to kick the ground with his hoof to tell his teacher about the number written on the patient, for example, if von Osten wrote the number three on the patient, Hans would hit his hoof on the ground three times, and the results of his reading of the numbers were correct as long as the number was less than ten.

With these successes, von Osten's enthusiasm grew and he tried to push his promising student forward by writing some of the most difficult mathematical questions about patience, and he managed to teach the horse to understand what mathematical symbols and signs mean, so Hans began to give correct answers to some issues such as roots and numeric fractions, Hans proved to be a good student.

"If Tuesday falls on the first day of the month, what will happen on Sunday next week?" Von Osten asked Hans's horse in front of the audience, and he replied with six hoof strikes on the ground: "Okay, what's the square root of sixteen?" And the answer is- four hoof strikes on the ground from Hans's horse, and the audience was even more impressed when von Osten told them that Hans's horse can also read names and letters, so one hoof on the ground means "a", two blows - "B" and so Further. So Hans could pronounce some of the names he knew and he was adept at solving some simple math problems like addition and subtraction and he can tell you exactly the day of the month !! Although he was sometimes wrong in his answers, 90% of the questions asked were correct, and by some estimates, Hans's athletic ability was equivalent to that of a fourteen-year-old boy.

As Hans's popularity grew, especially after the New York Times wrote about him on its front page, the suspicions of some scholars and researchers of the existence of the hoax also increased, and the German Education Authority requested independent research and research on Hans's scientific abilities. The commission for establishing the truth consisted of two zoologists, a psychologist, a horse breeder and a number of school teachers, as well as the director of the circus, and the commission conducted a comprehensive test of the abilities of the horse Hans, and in 1904 the commission concluded that there was nothing in the scientific abilities of the horse Hans. trick or trick and that his answers are encoded using special methods.

The commission then turned to Professor Oscar Fuenquist, a psychologist who had some ideas and plans to unravel the mystery of Hans' horse, and he erected a large tent to conduct tests inside it and banish any outside influences that might interfere with the horse's responses.

Hans answered well the questions of his teacher von Osten and gave precise and correct answers to the questions asked by him and others, but when Professor Oskar asks von Osten, Hans's teacher, to stay away from the test site, something strange happens, namely accuracy and correctness the horse's responses decrease slightly.

In the end, Professor Oscar came to two important conclusions: if the person asking the question himself does not know the answer to it, the accuracy of Hans's horse's answer drops to almost zero. He also found that by forcing a horse to ask questions that he could not answer, he constantly fell to the ground.

Professor Oscar continued his experiments with an emphasis on observing human interactions with Hans, and he quickly noticed that the manner of breathing or sitting or the expression on the face of a person asking a question to Hans involuntarily influences the horse's hoof strikes every time, for example, every time Hans hits the ground with his hoof and when he reaches the right one with his right decision or refrains from it.

In fact, the test result was that Hans 'horse had nothing to do with mathematics, but it probably gave answers based on the facial expression and involuntary movements of the person being asked, it looks like Hans' horse used an ability that it possesses many other animals, understand or communicate using body language or the facial expressions of others, which could be understood with its help. Through some kind of movement.

After Professor Oscar came to this conclusion, he personally tried to imitate the horse Hans, asking people to ask questions personally to him and responding with a kick on the ground, as does the horse Hans, and at the same time carefully studying the facial expression or movements of the person asking question, and to his liking, he noticed that the people asking the questions are unable to control their expressions when he arrives at the correct answer, even if they knew in advance that the respondent was basing his answer on the basis of these simple expressions and movements.

And in subsequent years, scientists came to the conclusion that there are many animals that are sensitive to the measures and movements that their owners consider, and today the term "Hans Smart effects" is used to describe the influence of idioms or signal disorders on the response they receive, whether then a person or an animal. Scientists later discovered that some animals, such as dogs and horses, can even recognize the beating of a human heart.

According to Professor von Osten, he was never convinced by Professor Oskar's findings, so he and his clever horse, Hans, continued to follow their math lessons and take public excursions throughout Germany that attracted large numbers of spectators and spectators, although poor Hans had no knowledge. mathematics and no inclination to learn German (according to the conclusion of Professor Oscar), his ingenious ability to laugh at the chins of people who gathered to watch his reviews, really deserved the title of smart Hans.

So the story ended here, no, as usual, von Osten died in 1909 and the ownership of Hans's horse passed to another man named Karl Karl, a jeweler from Bierfeld, and Herr Karl decided to continue giving horse lessons to Hans, but he went even further, he taught four other horses to count and read, and the great surprise of the scientists was that two of these horses were blind, correct vision?

Even Professor Oscar himself admitted that his conclusions were not beyond criticism or 100% correct, writing in an article in 1911: "Some say that the horse was taught during the tests to respond to our reactions and expressions, not allowing it to use its independent thinking , and that before that, the horse really knew how to count and count, and that she just got used to the bad habit of not making mistakes? "

Some still insist on Professor Oscar's theory that the horse gets its answers from the expressions that come up with the questioner, but what about blind horses? ʿThere is another necrosis, and if it is not approved by scientists, then this necrosis is led by researchers in the field of parasiology or the science of supernatural and supernatural abilities, and this necrosis suggests that some animals have the ability to telepathy remotely, that is, they read the thoughts of the corresponding person to get the right answer, and there are also some! Which parties do you believe, dear reader, and what do you think about it?

For more information on this topic, you can type the phrase "Clever Hans" or the phrase "Elberfeld horses" in any search engine and our Arab horse brethren will be able to experience raising their horses, especially since Hans' horse was an Arab Horse !!!!.

UFO riddles

The UFO phenomenon is one of the most exciting topics that hardly a day goes by until some newspapers and magazines around the world tell us exciting stories about it, but what is the reality of UFOs and does it really exist? No one knows for sure, there are those who are for its existence, while others categorically deny it, but this topic always has a big question mark, and there are also hundreds of opinions that cannot be explained logically.

Note: There is a lot of news and photos of spurious dishes circulating on the internet, don't believe everything you see or read.

Although UFO sightings have been reported since ancient times, the resurgence of interest in them in the middle of the last century and their similarity to dishes is associated with an American businessman from Idaho named Kenneth Arnold. in June 1947, Arnold was flying in his private jet over the Cascade Mountains when he saw nine radioactive disks in front of him. When asked by reporters what these objects look like, Arnold replied that they resemble plates of food when they slide on the water surface, and therefore the name "Flying Saucers" got into the press and in books and dictionaries, although the scientific designation for this phenomenon is Unidentified flying objects or, for short, UFOs.

Since then, these flying objects have been sighted several thousand times throughout the Earth and its atmosphere, and major countries have studied them and formed government and other civil committees to document these observations. Sightings of these UFOs date back to ancient times, and evidence of these UFOs can be found in many ancient books and antiquities. In the heritage of ancient peoples, we read that the gods always descended from the sky in powerful and mighty vehicles and made sounds or flickers, and often the gods mated with people and produced demigods to lead them to victory in wars and organize their life in the days of war. peace.in Indian heritage, for example, the history of the use of gods falling from the sky with their powerful Vimana vehicles during local wars is repeated

Famous people such as discoverer Christopher Columbus had the opportunity to see a flying saucer the day before he discovered America. At 10:00 on October 11, 1492, Columbus wrote in the Register of the Bishop of Santa Maria: "Today we saw a light that floats in space and disappears, only to appear after a while, and this has continued for long periods."

In August 1566, in Basel, Switzerland, residents of the city watched for a long time in the sky multi-colored black and white balls dancing among themselves, which soon disappeared. With the advent of the press and other media, it has become possible to learn a lot about observations of these phenomena at present, and perhaps the first press report to mention a "flying saucer" is what a Texas farmer saw when describing his vision of a flying object resembling saucer and moving very fast, in 1878. And in the period between 1896-1897, a wave of observations swept across the United States, which was reported by the then press, the strangest thing was that three peasants saw in Kansas on April 19, 1897, at a height of ten meters, the three of them saw a large body one hundred meters, resembling a cigar and slowly landing on a flock of sheep belonging to one of the peasants. The UFO disappeared shortly after its pioneers abducted a small cow whose dismembered body was found the next day a few miles from the farm.

In the second half of the twentieth century, new types of observation appeared, which included a new element, namely, a direct meeting between the pioneers of these bodies and between people, whether by persuasion or abduction, and the fame of one of these incidents reached such a degree that its hero George Adamski was officially invited to meet with the Pope, Queen of the Netherlands and Queen of France. George Adamski, a native of Bologna and America, was living in California when he first met an astronaut from Venus on November 20, 1952, when he and six friends who had the hobby of observing the sky with telescopes traveled to the California desert to observe a flying saucer, which a few days ago was repeatedly seen in the sky. In his book UFO Lands, Adamsky says that he had a strange feeling that he was the one who wanted to meet the pioneers of this dish among his friends, so he asked them to leave him alone for a while, and he actually watched the six friends with the ship's distances landed next to Adamski and the man in black.), The man who looked so much like the inhabitants of the Earth told him that he flew in from Venus after his people noticed atomic explosions on the surface of the Earth (the bombs in Hiroshima and Nagasaki were detonated seven years ago), and that Adamski must have send this message to the inhabitants of the Earth so that they turn to peace, not war. Adamsky's observations and his encounter with the Venusian astronaut remained highly controversial even after the death of a man in 1965, while some consider him the greatest charlatan in history, especially after the discovery that the surface of Venus is inflamed and filled with asphyxiant gases, which discourages an astronaut like described by Adamsky, to live in it, others believe that he once visited their planet with them, and that when he entered the atmospheric belt of the Earth, he saw a magnificent sight, consisting of billions of light particles surrounding the apparatus, which in their shape resembled flies, and the view reminded him of the huge Fireworks Arena. Surprisingly, American astronaut John Glenn confirmed this observation in the same detail during his first flight into space in 1962, seven years after the so-called Adamski flight! Was Adamski truthful in his claims?

The second incident is the New Hampshire incident and invalidated by Barney Hill, a United Nations postal worker, and his wife Betty. On the night of September 19, 1961, the two of them were returning from Canada to their home in New Hampshire in their own car, when Betty noticed a bright light in space that moved strangely and sometimes looked as if it was spinning by itself. As the light approached, Barney noticed that it looked like a large disc surrounded by window openings. when the UFO was 20 meters away from him, Barney noticed that several people were watching him through the windows. The couple heard a strange sound and felt numb and sleepy, and then, as soon as they heard the strange sound again and noticed that they were now 35 miles from where they first stopped, the couple ended the journey to discover when they arrived home that their clock was off and that the clock of the House indicated that they were two hours late, so this incident would have passed without a fuss if the couple had not started to feel strange kinds of ailments. Barney had strange neck pains and high blood pressure with strange sores on his skin, and Penny began to have terrible dreams, and when a group of belt-shaped boils appeared around Barney's body, the couple began to see doctors. During the treatment, the doctors drew attention to the history of the UFO sighting and the spouses were sent for psychotherapy, while examining the car, it was proved that there were several round areas with traces of radiation exposure. And when psychiatrists resorted to hypnotherapy, little by little the details of what had happened became clear, and the testimony of the spouses coincided in the most precise details, despite the hypnosis procedure for each of them separately, and in the testimony of the spouses it turned out that after they felt sleepy, their injected into a foreign body by a group of prosthetists and how they unsuccessfully tried to pull Betty's teeth out. The couple describe these people as people with large eyes extending to the sides of their faces, no nose, with a mouth that is a small hole without lips, and before the visit was over and Barney and Betty returned to their car, space visitors showed Barney a 3D map of some stars that are connected by lines of different thickness. Betty realized that these lines were the navigation routes they used during their travels, and before saying goodbye, the group leader told them that they would forget everything that happened to them as soon as they got back to their car. Oddly enough, the astrophysicist managed to find the right place for this map among thousands of stars under the globe, on the basis of which it was assumed that the source of these visitors was a star (Zeta Reticuli), located 37 light years from Earth !.

And it was not without folklore about UFOs from incidents that cause horror and fear in the souls of its viewers, including our third incident, which occurred in Hopkinsville, Kentucky, USA on August 21, 1955, when baby Billy returned home frightened and told his *parents, that he saw a large shiny object landing near their farm, and his statements were not taken seriously at first, and after about an hour the three kids were already sitting. This creature, only three feet tall, with elephant ears and bright eyes, was rapidly approaching the house when one of the family members began shooting at him, causing him to disappear for a moment to reveal another creature of the same kind from a window, another in a tree and the third on the roof of the house, which caused the family to have a strange siege of this kind, during which the trapped people fired at the creatures to no avail, as the shots seemed to deflect when they hit their bodies, and the only thing that disturbed these intruders. there was a bright light emitted from hand torches. Around midnight, the family managed to escape from the farm and everyone rushed in their cars to a nearby town, state police said. When police cars arrived at the family farm, some of them noticed that luminous objects from the house were flying into the sky, and one of them flew over the police car at high speed, causing a huge hum. The tension of witnesses to the Hawkensville incident prompted many authorities to examine it, including the US Air Force and the Blue Book Commission, and the commission's report that the incident could not be explained on any logical basis.*

There are thousands of UFO sightings around the world, and authorities and organizations have documented and analyzed these incidents, and it has been found that many of them can be interpreted as observations of natural phenomena such as auroras or meteorites, comets and other phenomena that people who they are seen, they are thought to be UFOs, and there are many false sightings whose owners seek fame, but despite all this, the people who see them think they are UFOs.
Theories about the origin of UFOs:

1. The first thing that comes to mind is that these UFO-piloted machines are from sane creatures (whether biologically or mechanically) and originating from one or several billion stars and planets of our vast universe, in which our planet is a place of distance from sea. so if we believe these are ideas, it will be easy for us to believe that some of these stars or planets are thousands of years old before civilization, giving it the ability to use energy (to cross the vast distances that separate them from us) which are still unknown.

2. The secret weapon theory and they say the freshness of a group of scientists who claim they have evidence that proves that these UFOs are nothing more than secret weapons tests developed by major countries and that these countries are spreading rumors about UFOs so that the barrier of fear does not prevent people from approaching them.

3. The theory of a parallel universe: an exciting theory of fiction, the owners of which say that there may be a civilization that shares life on this planet, but it is invisible to us, because we are limited by our five senses, which, according to scientists, are not enough to perceive what is happening on this planet. universe.as for example, a dog can pick up sounds that we cannot hear because they are beyond the vibration that our sense of hearing can perceive.

4. Dissenting theories: These are theories that try to find a logical explanation for many UFO sightings, and a report by the Condon Committee and the Writers Committee formed by the US government to study all UFO sightings tried to interpret them all as either bright Venus or ionized swamp gas collected in the upper atmosphere, or a group of light insects flying at a height, or !.

There are other theories, such as psychological theories, which are driven by the so-called collective obsession, as a result of the complexity of life and the numerous crises of the modern era, which forces a person to withdraw with his unconscious mind into unfounded fantasies, such as UFO sightings, and many other theories, such as as a theory of returning to the origins, one of the supporters of which is the great physicist Einstein.

The UFO topic is a large and complex topic, and there are thousands of strange and bizarre incidents that have occurred in various countries around the world, the most famous of which is the American incident in Roswell, where it is believed that a flying saucer with its crew crashed there in 1947, and the government The United States hid all facts about the incident and argued that what crashed there was a test balloon for the atmosphere, like the Flying Saucer that landed in Kuwait in the 1970s, and many others that this rush does not allow us to mention.

Children's stars .. The skull puzzled scientists

Since ancient times, there has been a legend spread by most peoples and peoples about aliens coming from heaven, alien creatures depicted in various forms, bodies and names - these are messengers of the gods, and others are angels, who often are between heaven and earth, often to spread good and justice among people, and sometimes to destroy and burn large cities from Noise and scientific disputes to the present day.

Like most discoveries, chance played a major role in the discovery of the skeletal remains of the Star Child, in 1930, an American married couple took their teenage daughter to visit their relatives in a city in Mexico about 100 miles from the capital Mexico City, located in the neighborhood full of caves and tunnels of abandoned old mine tunnels, and she wandered between them to learn their secrets. at the end of one of these old tunnels, a surprise awaited her. She didn't expect her, there were two skeletons lying side by side on the floor of the tunnel, one for an adult and one for a small child, the girl removed the dirt from two skeletons and then collected their parts and transported them with her to the United States. to stay with her until her death, when they diagnosed it as a skeleton belonging to a woman in her 20s and the mother of a little boy. They estimated his age at about five years, at the time of their death, experts estimated him at 900 years old, here everything is normal and consistent, but the difference of opinion appears when examining the skull of a small child, where there are three necrophages around:

The child had a craniofacial deformity such as hydrocephalus, hydrocephalus, Progeria, premature aging, or other birth defects.

The child underwent the process of forming bones in order to change their shape, attaching them from childhood in the way of changing the shape and size of the skull, a method that was known in the civilizations of South America and the civilization of the pharaohs of Egypt, especially among the priests, and there are many examples of this.

The third is supported by supporters of aliens who believe that the child is a hybrid that arose as a result of the marriage of an alien with an earthly woman.

For the first ectopia, some scientists who have examined the skull exclude its deformation as a result of a certain disease, because its shape and consistency seem to be normal, and not the result of a certain symptom, and for the second ectopia, all models that came from ancient civilizations show that the reconstruction of the Skull Shape is located in the region behind the head because it may be attached to the inability of the previous two treasures to explain the shape of the skull of the child of the stars, prompting the owners of the third treasure to insist on its authenticity, and they are based on other anomalous things revealed by laboratory examination of the skull. the skull of the child of the stars, including:

The size of the hollow cerebral skull is normal, it is equivalent to 1600 cubic centimeters of any larger by 200 cubic centimeters the size of the hollow cerebral human of an adult natural and the largest is 400 square centimeters of the size of the hollow cerebral normal child of the same age as the Success child.

The eye cavity is oval in shape, and its depth is much lower than that of a normal person, and the nerve of the eye connects to the eye from below, and not from behind, as in a normal person.

The shape of the ears looks normal, but scans show that they are larger than usual and that they are also deeper than usual.

The skull has no frontal pockets in the frontal bone.

The last examination of the skull in 2004 revealed an unknown type of fibers in the composition of the bones of the skull.

According to the theory of evolution, the difference between the size of the brain in each human species reaches 200 cubic centimeters, starting with the human hemohapilis and ending with Neanderthals and modern humans, and this poses a big question and mystery to scientists, since, according to the theory of evolution, the skull of the child of stars belongs to an unknown person ...

Unfortunately, the girl who discovered the skeletons in 1930 cleaned them up and exposed them to the sun, destroying her DNA and thus making the task of DNA testing of skeletons difficult and difficult. Even so, scientists performed four DNA analyzes, the last of which was in 2004, and tests showed that the woman whose skeleton was found in the tunnel next to the Star Child's skeleton is not his mother, and that the Star Child's mother is indigenous. American. Who is the woman whose remains were found with the child of the stars? Why were they hiding in such a dark, lonely tunnel, away from human eyes? How did they die and why? Mysterious questions remain unanswered, waiting for science to one day arrive at a sophisticated mechanism for collecting, analyzing and testing DNA from ancient skeletons. About the fact and personality of the star skull baby.

Kolagina .. The woman who stunned scientists

The woman who stunned scientists with her supernatural abilities and became the object of a long debate about the truth of these abilities, some say that she is just a fraud, while others believe that her abilities are real, she moved objects without touching them, distinguished colors without seeing them , and looked at the frog, making her heart stop beating, and she was filmed in several documentaries, performing experiments and scientific tests.

Since ancient times, many stories have been told about people who differ from the rest of humanity in supernatural abilities, such as predicting the future, healing the sick, moving things and flying through the air .. People have often attributed these abilities to black magic or religious miracles, but modern science that rises over the occult and recognizes only tangible material things, attributed these abilities to visual tricks, sleight of hand and tricks based on physics and chemistry, which are the simplest ways to explain supernatural abilities for which there is no logical and scientific explanation, like those possessed by Russian Nina Kulagin, which have been repeatedly tested and photographed, even becoming one of the strongest pieces of evidence for proponents of what is known as parapsychology.

Who is Nina to us?

She was born in Russia in 1927 and she was fourteen years old when the Nazi Germans invaded Russia and laid siege to the city of St. Petersburg (Leningrad), and like many Russian children, Kolagina, along with her father, brother and sister, joined the Red Army and was sent into a battle that lasted nine hundred days, during which conditions were terrible, in winter the temperature was sometimes tragic, without water, without electricity, in these difficult situations. Fourteen-year-old Nina fought on the front line and worked as a reporter in one of the Russian T-34 tanks, where her courage and valor manifested itself in battle until she was promoted to the rank of chief of staff, but her life in the army ended when she was injured in one of the battles, she was hospitalized until she was injured.

Discover the supernatural powers of Kulagina

Nina claimed that she always knew that she had abilities that set her apart from others and that she inherited them from her mother, and there are many stories of her ability to know what is in other people's pockets without looking into them, in addition to her ability to diagnose diseases, although she knows nothing about medicine and has not studied it, which strengthened her faith and confidence in her abilities. Over time, Nina learned to develop and control these abilities. In 1964, Nina suffered a nervous breakdown, after which she was admitted to the hospital, where she stunned the doctors. she spent most of her time hand knitting, reaching out to the basket of balls of knitting yarn and choosing the right color without looking at the inside of the basket. the next year, as she recovered from her illness, Nina agreed to undergo experiments and tests by scientists, who found that Nina had the ability to distinguish colors as soon as she touched them with her fingers without looking at them, and she had the ability to speed up wound healing, like only she puts her hand on them. Nina's ability to move things without touching them probably caught the attention of scientists the most, and also brought her fame when Nina sat at the table and moved things like a clock or a tin can, matches and salty food, and apparently , Nina's abilities were not always available, since the experiments that she conducted preceded hours of preparation and meditation, she told the scientists that she had to do something. By the end of the 1960s, the glory of Kolagina began to reach the West, and in 1968 her abilities were considered at the first congress of paracyclology, held in Moscow, which increased the curiosity of Western scientists and their desire to explore and test the abilities of colagina themselves, and they got this opportunity in 1970, when a group of American scientists managed to meet her in Moscow, and they vary greatly in size and shape, and move slowly and irregularly. He also admitted that before the experiment, they took several measures to make sure that Kulagina was not cheating in her performance, forced her to constantly change her place at the table, and she was well examined to make sure that she did not carry a magnetic stone or hidden threads. In other experiments, to further validate the test, the scientists placed objects not affected by magnets, such as a matchbox, inside a glass barrier to prevent all effects, such as air movement or invisible threads. One of the scientists who studied the state of Collagin said that she had the ability to show the letter "O" or "A" on photographic paper without touching it, as well as transfer parts of the images that are focused on her onto the photographic paper, and in one of the collagin tests separated the yolk from the egg whites in a vase six meters away.

Perhaps the most famous and powerful of the abilities of collagin, which surprised and curious scientists, is its ability to control the functions of certain organs in the bodies of humans and animals, in one of the filmed collagin experiments it made the frog's heartbeat sharply accelerate, and then slow it down to a complete stop, and in another experiment, she controlled the heartbeat In the last years of her life, Kolagina stunned TV viewers when she made a small red speck appear on the hands of a European journalist.

In fact, one of the bad aspects of the experiments and tests carried out on Kolagina is their impact on her health, and even many in Russia believe that this was the main cause of her early death in 1990, scientists noticed that the experiments strained the colagina a lot. , sometimes she got red spots on her hands, and sometimes fire flashed in her clothes against the background of shock and this could last half an hour. In recent years, she has lost her sense of taste, and pains in her arms and legs have become very severe, as well as a constant feeling of fatigue and dizziness. Before dying, she suffered a heart attack that nearly took her life, forcing doctors to ask her to stop performing her superhuman abilities.
In conclusion, it should be noted that there are many skeptics about Kolagina's abilities, both inside and outside Russia, who see what she did as visual tricks and gimmicks, as well as the use of small magnets or thin and transparent threads, and cite evidence of this during the long preparation period before each test, and that most of her experiments took place in an uncontrolled laboratory environment, such as her apartment and hotel rooms, as some argue that what Kolagina was doing is also claimed by Kolagina's skeptics that it was simply a means of communist propaganda by the KGB during the Cold War. Kolagina's supporters respond to these claims by the fact that Kolagina was well examined before each experiment and was forced to constantly change its place within the experiment perimeter and place glass and plastic insulators between themselves and the objects being moved, and also by the fact that many of its tests were carried out in a controlled laboratory environment inside Soviet universities and many of those who researched her Nobel Prize in Science.

The deadly epidemic of 1918 .. He made 50 million in 9 months ... The horror could happen at any moment.

While the belligerent armies fought fiercely in the heart of Europe during the First World War, there was a hidden and mortal enemy lurking in humanity, an enemy so small that the naked eye could not see it, but it is more ferocious than all modern weapons invented by man. it will invade the bodies of a third of the world's population and cause the death of many people. But the emergence of new strains of viruses, such as avian and swine flu, has reminded of the ghost of this horrific epidemic and has left many wondering if this could happen again.

At the height of World War I, a lot of blood was shed and new devastating weapons were used that the world had never known before, such as tanks, submarines, aircraft and chemical weapons, and for many in Europe they could not imagine anything worse than the war in which they were burning, but they were wrong, what happens after March 1918 will be something other than making a heavy guest of the war-weary old inhabitants of the continent. The disease spread several times as an epidemic, the last of which was the Russian epidemic of 1889, which left more than a million victims around the world, but in general, the flu was not considered a fatal disease, especially for young, strong and healthy people, the victims of the disease were mostly vulnerable small children and the elderly, but this optimistic view would later be known as the Spanish flu, even though Spain was not the epicenter of the disease. Unlike the Warring States, which imposed strict restrictions on the press, especially with regard to casualties and the size of the victims, the news of the disease that was broadcast by the World Press was mainly from Spain, leading many to believe that Spain was the source of the disease.

Researchers identify several possible areas for the first infection, such as China, the United States and Austria, but first the disease spread widely in the United States army camp in Kansas, and then new infections appeared in several other states of the United States before the disease crossed the Atlantic. with American soldiers heading to battlefronts in Europe, and from there it quickly spread thanks to the development of modes of transport and the movement of warring armies to take the lives of their victims around the world.

The epidemic spread in waves, the first wave occurred in March 1918 and the severity of the disease was moderate and very similar in symptoms to the symptoms of ordinary seasonal flu, the victims were mainly children and the elderly, while young and healthy men recovered and recovered quickly from the disease, but by August 1918 there was a second wave, during which, with their victims, And young and healthy men became more susceptible to illness and death due to it, in contrast to the common flu, which usually affects people with weak immune systems, as the symptoms of the disease and its complications changed markedly, one of the contemporaries of that period wrote: "one of the strange complications of the disease is bleeding from the mucous membranes of the body, especially from the lungs. The disease was so fatal that some patients died on the same day that and their symptoms, and there have been many stories of people who showed symptoms in the morning and died and looked like dropped in the evening, or others who suddenly fell to the ground on their way to work and died a few hours later. These tales of the deadly nature of the disease fueled an atmosphere of horror, people began to move away from crowded places, and some stores began to discourage customers from entering and asking them to write a paper with their requests and leave it in front of the store door, schools broke down, and some factories and the markets closed their doors. There were entire families who died or a large number of their members died, and many children were left orphans, and the funeral of the deceased took at best only a quarter of an hour and only some of the deceased's relatives were present, and there were many dead left without burial because the clergy and the gravediggers were sick or dead, but in battle the disease was greater than that of the Allies. Many of the soldiers believed to have died in the fighting were not victims of the weapons of war, but died of disease. According to statistics, one third of the world's population was infected with this disease, that is, about 500 million people (estimated in 1918), and 10-25% of those infected have died, that is, between 50 and 100 million people, which is equivalent to 3-6% of the population. Earth at that time, the percentage of those who died as a result of illness compared to the previous year. - 65 years old while mortality is high as a result of the usual seasonal flu among children and the elderly due to their weakened immunity, the epidemic has spread throughout the world and left victims in all countries, even the remote islands in the Pacific are not spared their plight. in the United States, the number of infected is 28% of the total population and the number of deaths is 675,000, in the world there are 80 deaths. - 90% of the population, and many of these terrible outbreaks are associated with the return of soldiers from the frontlines of battles in Europe carrying the virus, in addition to the failure to apply the correct medical procedures to cope with the epidemic, the most important of which is the quarantine of travelers suspected of the presence of a disease.

With the beginning of 1919, the epidemic began to decline and disappear quickly, some attributed this to the growing medical experience of combating the disease and taking precautions in most countries against its spread, while others believed that the real reason for the end of the epidemic lay in the influenza virus itself, since he gradually loses his strength and deadly qualities. Despite the heavy losses inflicted on humanity, people quickly forgot about the epidemic, and perhaps the end of the war and the beginning of a new peacetime played a big role in the fact that suddenly the memory of the disease was lost after the news of the disease overshadowed everything, even some the researchers called it the "forgotten epidemic", but with the resurgence of the avian flu virus and swine flu, an epidemic broke out in the epidemic. Laboratories and medical research centers around the world began to study the 1918 epidemic to try to find out which virus caused the disease, how it originated and how it was transmitted to the human body, the forms of infection and the effect of the virus on the body, and already scientists were able to recreate the virus according to samples taken from the body of an Inuit woman who contracted the virus by monkeys and died a few days later. The destruction of lung tissue was so great that the monkeys choked and died as a result of blood-soaked lungs, which are the same reasons that historians mentioned the symptoms and complications of influenza in 1918, and experience explained to doctors why most victims of the disease are strong immunologists. a secret that has not yet been revealed. Experience has shown that the destruction that was done to the tissues of the body was not actually caused by the disease virus, but is the result of a defect in the secretion of proteins that regulate the functioning of the immune system and lead to its overstimulation, so that its reaction causes the destruction of cells and body tissues. Doctors are currently studying how to develop new treatments for dangerous influenza viruses in the event of a new outbreak like a pandemic, and despite the lofty goal of these scientific trials, there are some objections out of fear that the leakage of some of the revived viruses will lead to a new epidemic and global catastrophe.

Influenza Virus Supplement
There has been a lot of talk about influenza viruses lately, so we decided to end this article with a simplified explanation of these viruses and their types to help the reader understand some of the medical terms that often circulate in newspapers and satellite TV. :

- influenza virus type (A): the most dangerous viruses of the disease and the most deadly in the human body, wild birds are considered the natural host of most types of these viruses and are usually transmitted from them to animals, such as domesticated birds and pigs, and from them to humans and cause fatal epidemics, strains of this virus differ in their severity within a single pattern, for example, the 1918 virus:

- H1N1, virus causing the 1918 epidemic and swine flu in 2009 (50-100 million deaths) - H2N2, the virus causing the Asian influenza epidemic in 1957 (1-4 million deaths) - H3N2, which caused the Hong Kong epidemic in 1968 - H5N1, the virus that causes avian influenza, very dangerous, but it did not turn into a pandemic after - H7N7, which affects people, birds, pigs, seals and horses, Netherlands, 2003 (89 infected, 1 death) - H1N2 infects people and pigs - H9N2, Three cases in Hong Kong between 1999 and 2003 and all were cured - H7N3, several cases from 2002 to 2007 and they all recovered - H3N2, two cases in 2004 and recovered - H10N7 , the virus is dangerous and there was only one human infection in Egypt in 2004

- Influenza virus type (B): this type contains only one species and only infects humans and is a less serious and common type (A) and does not become an epidemic.

- Influenza virus type (C): One type and less common than the previous two types, and usually affects children and has a low risk.

The Mystery of Sudden Death Syndrome in Sleep .. Superstitions and Scientific Interpretation

Some questions make people believe in superstition, especially when science cannot give them a clear and understandable explanation, and one of these questions is sudden death syndrome in sleep, the strangeness of this condition is caused by its victims, which are mostly healthy and strong young people who are not suspect any pathological symptoms, but do not wake up from sleep and do not turn into a lifeless body with a tinge of fear on their face, as if they saw a terrible sight before death, and I did a lot of superstitions about this and wrote many theories and scientific studies, but the mystery remains unresolved. Sudden and sudden death syndrome is a condition of sudden and sudden death that occurs during sleep and there is no convincing medical explanation for its occurrence as it affects healthy young people who are unaware of any serious symptoms. such as heart disease, which usually leads to sudden death, most of the victims are male and middle age. There are a few people who have gone through this condition but have experienced death, and they described their experience of waking up from sudden feelings of fear, paralysis and inability to move. and something pressed on their chest so hard that they almost choked, as well as the feeling that someone or an object was with them in the room, but they could not move.

This situation was first seen in the United States in the 1970s due to several sudden deaths among Asian Hmong refugees, and since then, over a hundred deaths have been reported for the same reason. In Asia, this case is old and well known among the Laos and the Philippines, especially among Hmong males, so scientists believe that genetic factors play a large role in their occurrence, and it is believed that 43 out of every 100 thousand people suffer from this case, and doctors trying to find logical explanations for the current situation, the most important drugs, epileptic diseases, acute inflammation of the pancreas. Sleep disorders, genetic diseases and other scientific causes, which, with all the plurality and variety of their true causes of the syndrome, remain a mystery unknown to most doctors and scientists.

Some folklore researchers attribute Sudden Death Syndrome to some superstitious beliefs that the Hmong strongly believe in and motivate their origin, they believe that there is a creature named Betty bat, depicted as a fairy or old witch, crouching on the chest of its victim and sending she has terrifying nightmares to kill them and has been hard ever since. However, this theory does not explain why these deaths do not occur in the same proportion and quantity among other peoples who believe in the myth of a spirit or an evil genie who attacks its victims in a dream, especially since this myth has been widespread since the ancients times and is known to different races and cultures in different forms and is mainly due to the fact that it has been common since ancient times. The Arabs call this state "Jat" or "Abu Lapid", because they believe that a genie with that name crouches down and beats against a person's chest.

The gates of hell opened in Siberia in 1908 .. The mystery of the Big Bang

Can you imagine, dear reader, the appearance of a second sun in the liver of the sky next to our current Sun, of course, many will consider the appearance of such a phenomenon as the sign of the hour, and this is exactly what the inhabitants of the Tonkin region of Russia thought about one morning in 1908, so that the poor have urgently sent their messengers to the nearest! And you can, dear reader, laugh at these people, but if you woke up to see what they saw that wonderful morning, you would feel sorry for them and forgive, or maybe even agree with what was going on. their opinion, that morning was extraordinary and extraordinary in every sense of the word. In the meantime, there is no need to worry about it".

The morning of June 30, 1908 was a normal morning in Siberia, the sky was clear blue, and the golden rays of the sun quietly streamed between the leaves and branches of the green Siberian forests, giving the Earth a beautiful golden color and spreading an atmosphere of joy and optimism in the souls of the few people who lived in these remote and isolated locations. It was about seven o'clock in the morning when he looked up and saw a bright line suddenly appear in the sky and split it in half, leaving behind a long tail of burning flame, which made him so hot that he almost ripped off his shirt. and for a moment it seemed to him that a fire broke out, and then suddenly majestically, as if it were the sound of huge stones breaking and falling from the sky, or as if thousands of cannons were simultaneously firing their shells. Mr. Semyonov and his wife covered their heads with their hands in panic and horror. then from the north winds of a very hot storm hit the wooden houses of the city, knocking out windows, breaking locks and setting fire to some of them. plantings.Mr Semenov was not the only person that morning who saw and felt the explosion. many people experienced the same sense of fear and horror as Mr. Semyonov, some of whom described what happened as a luminous UFO surrounded by fire and flames in the sky and moving northward for ten minutes before it exploded majestically. One of the witnesses next to this place slept with his brother in a tent in the open air, when they suddenly woke up feeling that the earth was trembling under them, and a hot wind and a storm uprooted their tent, one of the brothers described what he saw later: "when I raised my head, I saw a strange sight, the fire was not in the trees around us, but in the clouds, and our sun was shining as usual, but there was a second sun with it in the sky !! ".

Fortunately, the area where the explosion occurred directly above it was almost devoid of residents, so no losses or deaths were recorded in that terrible incident, which, if it happened over one of the major cities, would inevitably leave hundreds of thousands dead, because that the explosion was so powerful that it was seen and felt by tremors that accompanied people at a distance of 350 miles and caused unnecessarily light bulbs !.

Although it was strange what happened on that terrible morning of 1908 and although the Russian local press wrote about the explosion, the tsarist government, mired in Moscow's problems, did not bother to investigate the incident, and the turbulent events that Russia and Europe experienced at the turn of the 20th century may , forced the world to ignore what happened in the Siberian region of Tunguska. The government agreed after Keulk lured her in with precious metals that the explosion might have left behind and that could be used in the nascent Soviet industry. In 1927, after years of research and investigation and listening to eyewitnesses, the Kuhlk expedition arrived in the area of the explosion. in Tonkyushka, and the view they saw there was truly terrifying, there were completely burned land and forests over an area of more than fifty kilometers, the trees were charred and fell to the ground in the opposite direction. But the surprise that shocked the scientists was that they expected to find a crater or crater caused by an explosion at a point directly above it, but instead they found a small forest of trees that did not fall to the ground like those around it , but remained upright like electric poles after they lost all their leaves and branches, indicating that the explosion was caused by the explosion. Since no crater was found, the process of understanding what happened became more difficult than scientists at first glance thought and came to the conclusion after reading the testimony of witnesses that the exploded object was a large meteorite with a diameter of about sixty meters that penetrated the Earth's atmosphere and the process of friction and combustion led to a violent explosion, but as all the analyzes and tests that they conducted on the burnt remains at the scene of the explosion, they did not give a convincing and logical explanation for what happened.

Today there are several theories about the Great Siberian Explosion, perhaps the most generally accepted among scientists is that a meteorite with a diameter of about sixty meters entered the Earth's atmosphere on the morning of July 30, 1908 and burned up in the atmosphere at an altitude of about 7-10 kilometers above the Earth's surface, and its combustion may have generated a blast wave. - 15 megatons (1 megaton = 10,000 tons of high-explosive TNT) is equivalent to one third of the power of the Kaiser hydrogen bomb, which is the most powerful weapon ever created by humans, and scientists believe that the reason why the meteorite fragments did not fly in and hit the surface of the Earth lies in the fact that it consisted mainly of dust and dust.

There are other theories that are less acceptable for scientists trying to explain what happened, some of us believe that inside Niazi there are chemicals and metals that can be formed when a nuclear explosion burns, in other words, that a meteorite of this type will look more like a nuclear bomb or hydrogen natural process. Another theory claims that the explosion occurred as a result of the passage of a small black hole in the Earth (1), but what weakens this theory is the absence of a crater or crater in the area of the explosion, since no similar explosion occurred on the opposite side (that is, in the opposite point on the other side of the spherical body of the Earth). There are also theories of supporters of supernatural phenomena and supernatural events, which explain what happened in 1908 as an explosion of a flying saucer that flew from other worlds, where they argue that the weirdness of the explosion and the absence of any crater or funnel on Earth, in addition to the absence of meteorite fragments on the place of the mysterious incident of the American Roswell (2).

In addition to all the previous theories, there is the opinion of people who witnessed the incident, and the opinion of large groups of religious and believers around the world and different religions who believe that the explosion of tonkiuski is only a sign of the hour (two suns in the same sky!) And it is one of a number of signs and signs that have occurred in the modern era, and which shows that the day of reckoning and resurrection is just around the corner, or the lowest.

Today the Great Siberian Explosion is a daunting problem for many around the world, especially with the growing (unrealistic, but certainly probable) expectation that the Earth will collide with an asteroid or giant meteorite that could wipe out all life on Earth.

1 - Black holes are just the remains or ghosts of dead stars, and when some stars die after their energy is depleted (as will one day happen with our sun), their atoms merge with each other and shrink their mass until they turn into a small dark object that cannot be seen, but it has a lot of gravity to attract and absorb everything that is near them. Black holes are one of the strangest phenomena in the universe that has caught the attention of scientists in the modern era to study and understand them, and is much more complex than this simple and concise explanation we have written, so if you want, dear reader , to see more information about them, you can write black hole or (black hole) in any search engine.

2 - The American Roswell Incident: This is one of the most famous UFO incidents in the modern era, in which a flying saucer is believed to have crashed with its crew near the American city of Roswell in 1947 and the US government withheld all facts about the incident and claimed that what crashed was a test balloon, while UFO proponents argue that the US government has no official US map other than moons. The business of Russian is that it was photographed and published in the sixties and you can watch them today through the program (Google earth) and you will find a comment next to it to say: Don't ask, don't say, don't ask, don't tell anyone !! Area 51 and the alien bodies at the Roswell crash site were a key part of the story of the famous American Independence Day movie, which I think most of us have seen talking about the alien invasion of Earth.

Time Machine .. Is Time Travel Possible? In the meantime, there is no need to worry about it" .

An eccentric inventor, whose lover died in an accident, made a miracle machine for her, he traveled back in time in the hope of bringing her back to life, changing the events leading to her death, but he soon discovers that changing justice and fate is an impossible task , this fictional plot is taken from a famous novel that has been the subject of several films. Philosophers are cynical: how do I return to the distant past when I was not yet created, and how do I travel to the future when it has not yet happened? Good answer, but like most philosophers, the mystery only gets more complicated.

Time travel is a dream lived by many and has become a fertile theme for many science fiction novels, and perhaps the most famous in the field is The Time Machine, written by H.G. Wells in 1895 and revolving its the story of a young eccentric inventor whose lover died in an accident and made a miracle machine for her. Cinematography has found in this story and other time travel novels an interesting theme for many fictional films in which we observe how the hero from time to time moves in a complex and sophisticated mechanical machine, then in a simplified way, where the hero sleeps or falls somewhere to be in one time and another dimension, and although most of the stories are about time and time travel, Similar to the story of the owners of the cave mentioned in the Quran, where several people hide in a cave in which they mostly sleep and sleep for centuries and then wake up in the second century and time in the future, the people of Feteron are surprised and amazed about their costumes and old money, and in the Arab folk heritage there is a tale of a man who was killed.

For centuries, the topic of time travel was limited to novels and fairy tales, but at the turn of the 19th century, Albert Einstein published his theory of special relativity and attached it to general relativity (1) a few years later, and this theory turned many concepts and principles of physics and astronomy upside down, since thousands of years ago man was dealing with three specific spatial dimensions of relatively variable quantities. The closer the speed of an object to the speed of light, the slower time for it, that is, a person who travels in an advanced spacecraft at a speed of about the speed of light (186,000 miles per second) will pass time more slowly than a person who lives on Earth , and on this basis, time travel is theoretically possible. (2) What if the future doesn't like our space traveler? That if he finds Earth in a hundred years, and its population doubles, and its resources decrease, and wars and famine increase on it, he will inevitably regret his journey and want to go back in time, which is also possible in theory, and whatever the space traveler needs to make such a journey - it is to return to the past.

But since walking at the speed of light in the near future is an almost impossible task, the fastest human spacecraft did not exceed the speed of several kilometers per second, so scientists looked for other ways to travel in time and found their way into black holes (Black holes), which are ghosts dead stars, the mass of which crushed and condensed until it turned into very small particles. And some scientists believe that there is some kind of black hole called roller perforation, if you mind entering it and somehow managed to get through it without getting lost in its core, he will probably tell about this to another post and in another times, but this theory remains true ink on paper, because scientists do not yet know exactly what happens inside black holes, let alone travel inward.

There is another theory that suggests the use of cosmic strings or strings to travel through time, and scientists say that these filaments were left behind by the great cosmic explosion from which the universe is believed to have originated, that these filaments extend long distances throughout the universe, and that the intersection of the two of them together generates a lot of energy, which can lead to the curvature of "space-time" and thus allow the possibility of time travel.
But all these theories agree that the transition through different moments in time between the past and the future is possible, and scientists have proven this by measuring the time on spacecraft that made extraterrestrial flights, and comparing it with Earth's time, although the difference in time obtained by scientists, was very simple, given the slowness of modern spacecraft, it was enough to prove the validity of Einstein's theory of the relativity of time.

The topic of time travel was not only raised by physicists and astronomers and did not dwell on the problem of a method or technology that would allow a person to make such travel, but also went beyond them to other questions raised by a number of philosophers and the inability of scientists to provide an answer and a healing answer to them. one of the most outstanding because it doesn't even exist in the future. a really weird puzzle. Although astronomers and physicists are undoubtedly geniuses in the field of laws and mathematical equations requiring superintelligence, they faced helplessly this simple philosophical question, which was taken by philosophers as an argument and proof of the impossibility of time travel, and a team of philosophers asked another question, not less problematic and controversial than its predecessor, which is that if in the future a person can travel in time, But we have not yet shared the travelers who came from the future to visit our time ?.

Of course, scientists in favor of the idea of time travel did not sit silently and did not watch how philosophers shower them with a barrage of difficult questions, but tried to give some answers and solutions to these questions, the most outstanding of these solutions is the theory of the universe or parallel dimension, which suggests, that each dimension of time has its own events. Thus, the so-called "Grandfather Paradox" cannot happen, because a person can travel to another dimension of time, but will not be able to influence events or change them. As for travelers who came from the future, some scientists do not deny the possibility the fact that in our world there are people of the current era who came from the future, but they are hidden and do not advertise themselves, because the people of the future will be at a high level of progress and urbanization and will be more human and more responsible. Another hypothesis argues that the possibility of time travel will be unavailable if humanity does not have the technology to make such travel, that is, future travelers will not be able to come to our present time if we do not first open the time frame that allows them to make this journey ...

In practice, it seems that there is no way to travel in time in the near future, and only God knows when humanity will be able to achieve such a degree of development that will allow it to make such travels, but we must remember that we are all travelers, that time passes by us along the time guide called Age, which feeds the days and years, and it is the inevitable journey that you see, simply speaking poet. :
I want young people to come back someday .. So tell him what the grizzly did.

1- *Far from going into complex laws and mathematical equations that most people would not be able to understand, the theory of relativity simply means that everything in the universe is relative and nothing is absolute, the simplest example of this is when you are driving in a racing car and observing things outside, for example, behind trees, your eyes see that these trees are moving (in fact, you are moving).) What you will see in this case is that the other car is stationary and not moving because it is driving at the same speed as you, and you can even exchange a conversation with her driver, or a person standing on the street will see that she is moving and moving, and on this basis all the concepts of our life are relative, and the perception of it varies from person to person to the person, your way at 40,000 miles per second! .. Hold on to something so it doesn't fly out of speed !.*

Mysterious fires in Sicily .. A city on fire

Imagine, dear reader, that you are sitting in your house and watching a movie or a football match, and then suddenly and without preamble, the TV in front of you turns into a fiery flame, of course, you will panic and inevitably think that there was an electrical failure, and rush disconnect it from the house, but how will you feel and react when you stand in the middle of one of the rooms after making sure that the electricity is off, and then suddenly you see how one of the lights glows by itself and turns into a fireball !! Of course, she will say that this cannot be, but let me disappoint you and tell you the story of an Italian city about the loss of even superstitious events.

Canneto di Caronia is a small, quiet town, the name and location of which are rarely mentioned on the map of the island of Sicily, and whose ancient houses and simple agricultural inhabitants rarely attracted the attention of tourists and strangers, who passed by the Greeks, Romans, Arabs (1), French and the Spaniards .. But that all changed in one of the cold December days of 2004, when the forgotten depository city suddenly turned into a crowded place of electricians, firefighters, newspaper and satellite TV reporters, many scientists, as well as some clergy looking for miracles, and the population of all 39 people remained temporarily to settle in a hotel outside the city.

The strange events in the city began when a resident named Mr. Fazzanio sat quietly in his house and watched one of the TV programs, and then suddenly, without any preface, the fire on his TV turned into a burning ball of flame, and soon the same mysterious fires began to break out in other homes, and it seemed that an unknown magnetic force was drawing the fire to electrical appliances such as, Even the strangest and most bizarre is that these incidents continued even after the local electricity company completely cut off the city's power supply, and in the following days residents started talking about other strange things that were starting to happen in their homes, such as mobile phones exploding and burning on their own and the stream of fire from the water taps! And about items of home furniture that creep away from sources of electricity, as if there is a latent force that pushes them and moves them in the opposite direction from everything that has electricity.

Gradually, life in the city of de Caronia turned into a terrible nightmare, and some residents began to whisper among themselves, believing that genies and demons lived in their homes. Landline telephones rang at night, but no one spoke when the phone was answered, and the automatic locks of the car doors were locked and opened themselves, even the firefighters who broke into the city could not find a logical explanation for what was happening. Electricians made sure that the wires and electrical connections were in good condition and they could not find any defect or anomaly in the small wire network that powers the city with electricity, even scientists and researchers who came to the city after hearing about the mysterious events in the media have not given any clear scientific explanation for what is happening. One hypothesis was that the fires were caused by volcanic activity in northern Sicily and that this activity led to the emission of electromagnetic charges that reacted with ordinary electrical current and caused fires in electrical appliances, but this hypothesis did not explain why these alleged volcanic emissions were concentrated only in the town of Di Keronia and did not spread to nearby towns.

There is a second hypothesis, which states that the cause of the fires is a device that generates an electromagnetic field that releases energy charges in the form of electric detonators, which, according to the hypothesis, led to fires in different parts of the city, but this hypothesis did not explain how residents did not hear thunderous sounds that look like explosions. - 15 kilowatts could not be artificial, but they were inevitably caused by the landing of a flying saucer near the city, creating a massive magnetic field that caused the fires, but of course there is no physical evidence to support this hypothesis.

The theory or hypothesis most accepted by researchers is that a fire broke out in the city due to the accumulation of a large amount of static electrical charges, that is, the kind of charges caused by rubbing a plastic comb with human hair or that we observe when removing and wearing clothes from nylon or polyester, which generate small electrical sparks that can be easily seen inside.

Over the next months, electrical engineers and workers wrapped all electrical wiring and connections in the city with a special type of plastic film to prevent contact with electromagnetic charges.in April 2005, the mysterious fires stopped and residents returned to their homes. although the phenomenon has stopped, scientists have still not been able to figure out the true cause of the fire in the city and why there were a large number of static electric charges in its vicinity. some blamed the railway company for laying a railway line near the city, but there was no scientific evidence to support this view.

And since science and modern technology could not identify a clear cause of fires in the city, for a start, some residents tend to believe another hypothesis, since people have used it from the very beginning of creation to this day to explain things that they are not able to understand and find. her explanation is a hypothesis about a human being in a house or in a place where a fire is burning.

1-Muslims (Arabs, Amazighs) attacked Sicily from the 7th century AD during their battles with the Byzantines, and then ruled the island and parts of southern Italy for the 10th century, and at the end of the 11th century AD, their rule ended and they lived for a century under the rule of Christian kings, and then were forced to leave the island and leave it. Muslims have had a great influence on the culture and heritage of the island, they developed methods of irrigation and agriculture and were the first to bring and plant citrus fruits such as lemon and orange, for which Sicily is famous, as the Sicilian language has about three hundred words derived from Arabic words in mostly related to agriculture and food, as well as the word (mafia).)

Find a strange creature in Mexico ..
Scientists believe it is an object that came
from space! In the meantime, there is no
need to worry about it" .

A strange alien creature appears in Mexico! Is this a new trick that failed media outlets are trying to get people's attention with? Maybe .. But the news also spread on some sober and respectable websites and newspapers and there is a short film broadcast on Mexican television showing a group of scientists examining the remains of this creature that resembles those we see in science fiction films, and although this topic is hard to

believe, but we decided to translate it as published by the Daily Telegraph and attached a short film to it.
Is this strange creature really an alien child who came from another world, or is it part of a carefully planned hoax - and was his death the cause of a mysterious vendetta?

Mexican television aired in 2007 the arduous story of a farmer finding a strange "alien" child living on his farm. The farmer was so scared to see this abnormal creature that he drowned him in a water channel inside his land, and now, two years after the accident, scientists were finally able to announce the results of their tests on the body of an insidious creature.

At the end of last year, farmer Maro Lopez handed over the body of the strange creature to scientists at the University of Mexico, who conducted several tests on the body, and also checked its DNA.

Farmer Lopez claimed to have found the creature on his farm and that it took three attempts to drown him to kill him, during which he kept him underwater for several hours.

The results of tests carried out by scientists showed that the creature is not known and not scientifically classified and that scientists have not seen anything like it before, the characteristics of its skeleton were similar to the structure of lizards, but its teeth differed from human teeth in that they were without roots, and scientists came to the conclusion that in the creature, perhaps the only similarity between the creature and the person is limited to some joints in its body, which are similar to those found in humans. The creature's brain size was very large, especially the back, which led scientists to conclude that it was highly intelligent.

This creature confused the scientists and overwhelmed them. Even stranger is the mysterious death of Farmer Lopez a few months after he killed the creature.

According to 32-year-old American flying saucer expert Joshua Warren, farmer Lopez burned to death in his car, parked on the side of the road, and the heat generated by the fire and the flames that turned the car into ash was too high compared to that generated by natural fire. ... Which led many experts and people to believe that the alien's parents killed Farmer Lopez out of revenge.

There are several sightings of flying saucers and the Crop Circle phenomenon (1) in the area where a creature was found that may have been abandoned or forgotten there by aliens.
Mexican flying saucer expert Jimmy Mozen (56) was the first to publish this story and claimed that it was true and not a hoax, and that farmer Lopez had assured him before he died that there was another creature similar to the one that drowned him, but escaped when he stepped towards him.

The mystery of the mysterious creature aroused the interest of many readers who expressed different opinions on this topic, some wondered why aliens leave their young *children on Earth? While one reader was wondering why cosmic beings don't wear clothes (the creature was found naked) !.*

The story of a strange creature has spread so widely lately that it has reached far beyond South Korea and China !. Mexico isn't the only region where strange events have taken place, and Germany, for example, saw 366 reports of flying saucers this year !.

1-strange circles and shapes appear all over the world in crop fields like wheat or corn, and some people believe they are due to flying saucers, while scientists say it's just a trick performed by some with the purpose of attracting attention.

Superpower Man .. Fact or Fiction!

Electricity is deadly .. This is what we learned from childhood and it has become a truism in our life, and we cannot imagine that a sane person would hold electrical wiring at home with his bare hands, because it means that exposure to a strong electric shock can sometimes lead to death. But have you ever wondered if there are people who do not influence them with electricity? Science says no, all people obey the same laws of physics, and superhuman beings exist only in science fiction and science fiction films. But are there people who defied the laws of science and confused Reality with fiction?

We all have an electrical charge on our bodies, but we do not feel it or pay any attention to it, because it does not cause us any problems, except for some slight bites that we can sometimes feel when we hold the doorknob or any other metallic object, and this bite is the result of the discharge of static electrical charge accumulated on our bodies, and we deal with sensitive electronic devices while holding any metallic object. In addition to static electricity, we deal daily with alternating household electric current in all joints of our life, and we learned from an early age that this type of electricity causes painful shock and can lead to death on contact with it, so we are careful in our homes. to cover and block all electrical wires and make sure the electricity is useful .. But they are also dangerous and sometimes fatal to people, and this applies to all people, except for a few people around the world who claim that electricity does not affect them and defy all the laws of physics by manipulating deadly electrical wires in front of camera lenses without batting an eye.

Jose Rafael Alaya is one of those people who baffle scientists, he is an ordinary forty-year-old man living alone in a small rented room in Puerto Rico, and at first glance Jose seems like most people who spend their whole lives on the sidelines, come into the world and leave him without feeling them and not paying any attention, but Jose has a special talent in his small workshop that specializes in the repair of electrical appliances. Although Jose appears in front of the camera lens and demonstrates his abilities several times, scientists find no explanation for his lack of electricity and believe that he is either a superhuman doing miracles, or it all has to do with some kind of trick.

And in England there is a grandmother named Mavis Bryce, who is sixty years old and who also claims to have extraordinary electrical abilities, but unlike Jose, her imaginary abilities turned for her more a curse than a blessing, this old woman claims that for For four decades, she donated dozens of electrical appliances and burned hundreds of light bulbs, just touching any electrician's hand is enough.They sometimes damage sensitive electronic and electrical parts, but do not damage the refrigerator or TV. Ms Pryce says that what has eased her financial losses due to her extraordinary electrical capabilities is the guarantee system offered by electrical appliance stores to their customers, which allows them to return faulty appliances within a specified period of time.

Superpower Man .. Fact or Fiction! In the meantime, there is no need to worry about it ".
Old Man Zhang Demonstrates His Abilities

In China, there is a 71-year old communist who was raised by the people to the rank of Prophet! It is Zhang Dik Gai who has puzzled scientists with his ability to generate electricity from his body! He can light a light bulb, put it on his head or ear, and he can fry fish in the palm of his hand in two minutes! How he does it? He doesn't know himself! He says that a 240-volt electric current passes through his body, and although scientists say that this statement in particular is impossible, Zhang answers them coldly, challenging to prove otherwise, and in 1994 they put him to the test in one of the Chinese academies , but could not find any explanation for his abilities and, finally, they were illnesses of his friends and relatives and people began to call him "electric man".

Another supernatural manipulator of electricity - Slavis batets from Serbia, this strange man who looks like he came from another world, manipulates deadly electrical wires like he manipulates matches, and claims that he can withstand an electric shock of 20,000 volts. and he, like others, did not know about his extraordinary ability only by chance, he and one of his friends are his relatives. And already appearing in several documentary short films, the photographer proves his abilities defining the laws of science and physics, and even went further to defy death itself as they chose to sit on the killer's electroshock and electroshock chair.

In conclusion, it should be recalled that there are many similar stories about people with superhuman abilities for electricity, but we must not forget that electricity is deadly and that even if these stories are true, they are an exception and an anomaly to the general rule, therefore, please don't be reckless, dear reader, and try your luck with electrical wiring.

Problems of the human mind .. Elixir of life and the search for immortality

I love old photographs, especially those older than a century, I search the Internet, I look at it carefully, I look at it in the presence of men, women and children who appear on them .. The way they stand and sit .. Their facial expressions ... Costumes and clothes .. Always wondered how they live and where they die! The beginning and the end always attract me .. But it also confuses me: between the beginning and the end, life can be long and short, but it is inevitable with one inevitable end .. The end that people wanted to avoid since ancient times .. Or, by at least to postpone .. Does the science that has occupied the minds of people for centuries succeed in this? Can a person really live for two or three centuries in the future? And will scientists be able to finally create the elixir of life, which their ancient peers were looking for in vain?

Human life begins with a scene that the mind cannot imagine .. The great marathon in which millions of sperm are looking for one end .. It is surprising that this magnificent mass race does not receive any media coverage, does not write newspapers about it and does not convey reality by satellite TV !! In addition, the only winner of it is not awarded any medal or awarded any medal, despite his great achievement !. Its only reward will be to enter a spherical body called an "egg", and without it, the door will close in the face of millions of lost racers who will be left to die slowly outside.

The process of combining the winning sperm with the egg is the first step in shaping a new life. The first step on a thousand-mile journey of life to its inevitable end. It's funny that some critics of the world, like our poet and philosopher Abu Alaa al-Ari, curse this first step and consider it a crime committed by their parents against them !! :

This is what my father did to me, and what I did to everyone.

However, regardless of whether we are happy with our arrival in this world, human life, like the life of all creatures on our Blue Planet, goes through many stages, but they all end in one end inevitably, and the time interval between the beginning and the end we call it the name of age, and it lengthens, shortens and changes ..

Some insects live only a few hours or days. A mouse can live at best two years. A dog is 13 years old, a cat 15 years, a horse is 30 years old, and an elephant 70 years old. A human can live 80 years. Some species of turtles are already 190 years old. New Zealand liana Tuatara. can live 200 years. Bowhead whales can live for 220 years. Brightly colored Japanese koi fish can live for 245 years.

Some species of molluscs live more than 400 years. Some species of willows (bristle pine) can be up to 5,000 years. A bacteria called bacillus permians is 250 million years old

!! Scientists in the United States were able to bring it back to life in 2008 after it was extracted from a sodium chloride rock found in a cave in Mexico.

But is there a being that does not die?

The answer is yes and no right away !! .. There is a strange jellyfish called Hydrozone and its scientific name (Turritopsis nutricula), this little sea creature is the only creature on earth that cheated death and managed to cheat it. When this animal reaches the stage of senescence, which is usually followed by death in all creatures , this "rogue" jellyfish changes its life stages to return to its primary stages, a stage known as a polyp, and then begins to grow again until it reaches old age, and so on. This complex process can continue and repeat endlessly in these wonderful jellyfish, that is, they are "immortal" in every sense, but this does not mean that they do not die, this organism often preys on other organisms and therefore dies just like others. Scientists who have studied this jellyfish say that it uses a certain mechanism to recycle its life, transforming its body cells from one type and shape to another, a mechanism called transdifferentiation, and can be observed on a limited scale in some reptile species, such as the salamander. , which can compensate and repair some limbs of its body if they are damaged or lost.

But what about a person ?. How long can he live? And can he become immortal?

The funny thing is that among some people there is a belief that the age of the ancients was much longer than the age of people today, although the average age of a person in antiquity would not exceed 25 years at best, and in the Middle Ages this figure dropped to 18 years as a result of epidemics , hunger and wars.

Prophet Noah lived for about a thousand years, while other Old Testament prophets lived for hundreds of years. The Quran tells the story of the cave owners who lived in their cave for 300 years. There are also those who were resurrected by God, such as Al-Azir and his donkey, a hundred years after their death, and these stories are months before we go into details in this rush.

In our time, the average age of a person has increased as a result of the development of medicine and the availability of food, so the number of centenarians has also increased. newspapers now occasionally read reports of elderly people, but what is lacking, especially in third world countries, is documents proving the true age of these supposedly centenarians, since most of them do not have an official birth certificate. Due to this lack of evidence, most modern architects are either American, European, or Japanese, because their countries have a long history of documenting the births and deaths of their populations. According to official records, the longest human life in our time was recorded by Frenchwoman Jeanne Calment, who died in 1997 at the age of 122 years and 164 days. This Frenchwoman lived a long life and experienced the events of two centuries, during which she lost her husband, only daughter, only grandson and all the relatives and friends she knew to be alone in a nursing home. The funniest thing, dear reader, is that Jane Calment lived alone and without anyone's help and care until she was 110 years old, rode a bicycle until she was 100, smoked cigarettes from the age of 21 and did not leave this bad habit until she was 117 !!.

Returning to our question about the possibility of becoming immortal, in fact it has not been proven that there is a person who managed to overcome death only in Legends, as well as in some stories of religious heritage, and the story of al-Khader is one example of this, although stories about He is contradictory, Muslims differ in that he is a prophet or simply there are many stories about the reason for the longevity of the Greens, perhaps the most famous are his stories with the eyes of life, which return where the greens came from to reach this eye and in two centuries.

In mythology, Gilgamesh was the Sumerian king of Uruk (2750 BC), the first seeker of immortality, and his story was immortalized in the long poetic epic that bears his name. Although this story is just a legend, it carries wonderful meanings and judgments about the meaning of life, death and immortality.

Where is Gilgamesh going? The life you seek will not find it. When the gods created humans, they made death their share and locked life in their hands. You Gilgamesh, fill your belly. Stretch it day and night. Make every day a holiday. Dance Lachia night and day. Risk of light clean clothes. Wash your hair and take a dip in the water. Pamper your little one holding your hand and enjoy holding you in his arms. This is the share of people ... (Epic of Gilgamesh / tenth tablet / third column)

These verses are recited by Sidori, the servant of the goddess, when she tries to convince Gilgamesh that death is the inevitable fate of mankind ... there is no escape from her ... Therefore, instead of wasting time looking for the secret of immortality, I advised him to make "every day a holiday" and enjoy life as much as possible before he dies .. Because such is the fate of people. But Gilgamesh ignored the advice of the beautiful barmaid and continued on his way through the waters of death in search of the coveted vagrant. The magical herb that gives eternal life, which he uprooted from the depths of the sea in Dilmun (modern Bahrain), was swallowed by a snake while swimming in the river. In the end, the hero sadly returned to the Uruk, but when the ship carrying him approached the walls of the huge city and the courtyard he had built, Gilgamesh realized that such a great work as the wall would forever immortalize his name.

If Gilgamesh found the secret of immortality in the wall of his city, then the ancient Egyptians gave the world the greatest structures that have amazed humanity for centuries in order to reach this eternity. Unfortunately, most people do not understand anything about the meaning and philosophy of the life and death of the pharaohs, and some wonder what it means to waste all those enormous wealth and resources of the ancient Egyptian kingdom on the construction of grandiose stone tombs, such as the pyramids .. But most of them do not understand that the pharaohs built and built not for death, but for life, because death was in their minds not the end, but a new beginning of another Immortal Life, and a deeper study of the doctrine of death among the ancient Egyptians will reveal that this is the most ancient human doctrine of the concept souls.,

In the Middle Ages, the myth of the "elixir of life", which gives immortality and eternal life, spread, and although man never invented such a magic elixir, the search for it was of great use in the advancement and development of the science of chemistry, where the manufacture of this elixir became the target of many experiments and tests. especially in the Golden Age of Humanity.

The failure of science and medicine to prolong life and delay aging has led many people to resort to witchcraft and witchcraft to achieve this goal. Magical rituals associated with the return of youth were often bloody and cruel, for example, the Hungarian Countess Elizabeth Bathory cut the arteries of her unfortunate maids and left them hanging in the bathtub to bleed to the last drop, and after the death of the poor girl, the Countess took out the body and bathed in her blood because she believed that this blood had magical properties. Another example is the story of the Spanish woman Enrica Martí, who abducted poor children from the streets of Barcelona to kill them, and extracted fat from their bodies to use it to make magic recipes for eliminating old age, which she sold for large sums to the benevolent daughters of the city from the noble and high class.

These days we hear a lot about the so-called red Mercury, which some claim has amazing properties, including longevity, a lie that many gullible people spent huge sums of money to get at least a little of this magical substance, and which still exists. so far not proven in the first place.

Why are we getting old?
Far from myths and magic recipes .. Aging is wreaking havoc on the global economy because of the retirement benefits governments have to pay to older former workers, as well as the costs of treating and caring for the elderly, which are growing day by day as a result of advances in medicine and remedies. In addition, aging deprives humanity of its best scientists and thinkers, who have reached the pinnacle of their creativity and intellectual maturity as a result of the accumulation of experience and experience over the years. For these cumulative reasons, eliminating aging or reducing its negative effects would be of immense benefit to all of humanity.
However, attempts to prolong human life and slow down aging require first of all knowledge of the causes of human aging, and scientists, despite long-term research and significant progress in this area, have not yet settled on a comprehensive theory explaining why aging occurs. But in general, aging occurs at the cellular level (Cellular aging), when the cell stops dividing and regenerating after a certain number of divisions, which is associated with a defect that occurs in the process of copying DNA molecules (genetic code) during division, therefore, the greater the number of divisions, the more imbalance and deficiency in the DNA code of new cells also increases, which leads to an increase in the number of divisions.

Unlike healthy cells, cancer cells can completely copy DNA molecules, allowing them to divide indefinitely. This process, that is, stopping healthy cells and continuing to divide cancer cells, does not happen on its own. scientists believe that certain genes and enzymes influence this process. they hope that by identifying these genes and enzymes, they can slow down the aging process. for example, modifying these genes will allow healthy cells to continue dividing, thereby delaying aging while stopping harmful cancer cells from dividing.

Of course, there are many other theories about aging that take a long time to delve into, but the general structure of these theories links aging to genetics and genes, which act as an internal biological clock that determines the life cycle of each organism and maximum age.

Can scientists make the "elixir of life"?
As mentioned above, scientists are actively researching genes involved in the aging process. In fact, scientists really managed to extend a person's life, the average age of a

person a century or two ago would not have exceeded thirty years, mortality was very high, and it was not unusual for people to die in their prime. Today, with human life in the West, eighty years, and I really did not know humanity in its entire long history, a large number of elderly users, as in our days. People shake their heads sadly when they hear that someone died in their fourth or fifth decade, saying that they are too young to die. Today, serious research is underway to extend human life using genetic engineering, hormones and enzymes, and scientists have managed to achieve amazing results in some of their experiments on laboratory mice, where they were able to extend their life up to 40%, that is, if these experiments on humans are successful , then the average age will rise to 120 years, and there are even some scientists! And this is through certain hormones that control the functioning of genes.

And if a person managed to extend his life by one or two times in a century, then imagine, dear reader, the future prospects of human life in the next century or two, today scientists no longer talk about adding only a few decades .. No, he is aging ! But here and there we hear whispers about the summoning of death itself, and ideas and theories have already appeared on how to make a person immortal forever, some of which may seem to us today as unimaginable as our ancestors believed the car, plane and computer were contrived science fiction ... Or, as we were only thirty years ago, we do not even imagine, even in a dream, that the day will come when we will hold the phone and talk to it, walking down the street and not dragging hundreds of meters long wires behind us. ... These things were only seen in science fiction and spy films and in a naive way that did not match modern cell phone technology.

Perhaps the most recent proposals for human health and life are those that seek to use nanotechnology to create micro-robots that can easily pass through blood vessels to find viruses and microbes that cause diseases, to kill them instantly, and to promote recovery. and restoration of damaged tissues.

There is also an idea or theory called uploading consciousness into a computer, which believes that in the future people will reach such a stage of scientific development that they can completely copy their memory from the brain to a computer or to an electronic chip. After all, a clone is the person who over and over again loads the memory of his past life into his brain at birth or in the first years of his new life and restores all his previous experiences and, of course, the child or someone like that would not need go to school and learn sports, drive cars, communicate with people and the law .. Because all this will be stored in his brain, and even such a person can turn into a genius, because he has so much experience .. Dear reader, imagine ourselves, that we can clone Einstein and other geniuses several times, while preserving their full memory .. What will the world be like then? ..

In fact, in the modern world there are many discoveries, theories and hypotheses about the extension of human life, and there are people who believe that in the near future, humanity will be able to achieve much that may seem impossible to us now, so we hear about people freezing their bodies in ice after death in the hope that scientists can bring them back to life in the future, a technique called Cryonics, which in America costs $ 880,000 to freeze just the head and $ 1,150,000 to freeze the entire body. Although this method is relatively old, only two hundred people actually froze their bodies between 1962 and 2010.

In conclusion, it should be noted that everything that we have mentioned in this article about extending human life or making it immortal are just theories that one day may or may not come true. Who knows ? But will we survive if I check ?. - I doubt it. There are people in their fifth or sixth decade, if you subtract from their age hours of sleep and work and their sad, gloomy and difficult times, then there will be no more than on the fingers of years. Then what good is a long life in the Arab world? What good is it to live a hundred years in grief and sorrow over missed and missed opportunities .. Oh God, how much I weigh !!.

When we say that the average age of a dog is 13, this does not mean that there are no dogs that die at five or live until they reach 25, but rather that the indicator here is the age at which the dog should live. ...

Where have the bees gone? .. Is this really the end of humanity?

One of the letters I recently found talked about the mystery of the disappearance of bees and its connection with the end of the world, and I, as always, thought it was all fiction .. But after doing a little research on the Internet, I found that in this actually there are real measurements .. Scientists are really puzzled .. They think .. They are surprised .. Why did the bees do this? .. And where will he be ?!
On a sunny spring morning, Albert Einstein and his wife sat in the garden of their house and ate breakfast, leisurely sipping their morning coffee, their eyes following with interest the headlines scattered on the table in front of them.

Suddenly, there was a familiar buzzing in the air, and a little Bee appeared, circling persistently around Einstein's thick white hair, and the brilliant scientist raised his head and told his wife, looking at the Bee. :

"If the bees disappear from the face of the earth, a person will only have four years to live!"

For a moment this phrase seemed strange, but in fact there was a lot of wisdom and foresight in it, how not to come from someone who has always been called the smartest person that humanity has ever known.

But the only problem with this wonderful sentence is that Einstein never said it !!.

Yes, dear reader .. This sentence, which is widespread on websites these days, was not spoken or released by Einstein, this man was a genius in mathematics and physics .. But it hardly has anything to do with bees neither near, not far.

Bees are very important to humans, not only because of their honey and wax, but also because these persistent insects contribute to the pollination of a large number of plants that humans need for food and life. For example, of the 24,000 plants living in America, three quarters depend on bees, insects and birds for pollination.

Now imagine what would happen on Earth if the bees disappeared ... Of course, most of the plant species that rely on these flying insects for pollination and reproduction will become extinct, leading to the death and extinction of many animals that rely on the plant for their food .. Of course, the disappearance of all these plants and animals will have catastrophic and destructive consequences for human life.

But are bees really disappearing?

Yes ... It disappears, and strangely puzzled scientists overwhelm them .. Over the past five years, the number of bees in large areas of the world has decreased by 30-50% and continues to increase. Scientists have dubbed this strange phenomenon Colony Collapse Disorder (CCD).

The curious thing about this phenomenon is that the bee colony disappears suddenly and almost without warning! Beekeepers around the world wake up in the morning and find their bees empty on their thrones .. The whole bee disappeared without any signs or symptoms of disease, invasion or disruption of the hive .. The worker bees simply left the Hive and never returned to it.

And the really puzzling bees that leave their eggs in the hive .. Under normal conditions, bees do not leave their hive until the eggs hatch. In addition, the food supply in the abandoned hive is filled with honey and pollen collected by the bees for food .. Safe and clean warehouses have not been plundered by invading bees or vandalized by insects attacking apiaries such as wax moth and honey beetle.

The strangest thing is the survival of the queen bee in the abandoned hive .. Sometimes the queen dies and the bees leave the hive for this reason, but what distinguishes the CCD is the stay of the queen in the hive and the disappearance of the workers.

In short, it's like entering a house and finding that the furniture and the floor are completely clean, the food is still hot on the tables, the fire is burning in the ovens, the news is on TV .. But the people in the House are not visible .. They disappeared and never returned ! ..

But why do bees disappear?

In fact, as noted earlier, scientists in great bewilderment about the reasons for the collapse of bee colonies, many theories have appeared that tried to explain the mystery of the phenomenon attributed to different worlds and diverse:

- Various viral diseases affecting bees; - malnutrition as a result of the reduction of green spaces; - global warming and its known impact on the environment; - immunodeficiency and some types of mites and fungi; - side effects of some medicines used by beekeepers to treat bees; - as well as electromagnetic waves and vibrations from various devices used by humans, including a cell phone, which, according to some scientists, affects the ability of bees to determine the way back to the hive.

But, despite all these theories and hypotheses, the mystery remains, and there is still no clear final explanation for the phenomenon of the collapse of bee colonies.

Self-burning human

We are now in one of the famous Andes caves, namely in the province of Pasco in the Republic of Peru. The scene of the two Spanish archaeologists who came here to unearth the relics of the ancient Inca civilization that are rooted in the depths of history were respectively de Toledo, an old man of seventy, and his enthusiastic student Pizarro, a young man at the end of the third decade, as well as a group of miners. and Pizarro shouted at the workers with an ardent voice while his teacher interfered, before being resumed again with another research trip and so on.

It was when one of the workers shouted in the local language (Quechua), and of course Pizarro just went to him, where the man brought out a small statue covered in dust ...

It was a small figurine representing the upper half of the face of a grotesque man, as if he had come from the depths of hell in a Strange-looking necklace (Pesaro) at the moment when they were available with the mysterious glow of the statue's eyes, but he did not care what consolation from the dust lingering in the air as a result of a party .. Several words were engraved on the necklace that he tried so hard to learn or read that he could not: they seemed to be written in a language he had never studied in his life, although he had deeply studied all the local languages of the ancient Hispanic people, but it was a really strange language .. Then he stood at the side of his teacher, holding the statue that his teacher received and examined it, and then did not follow this pole, his eyebrows said behmm: a strange statue does not die for civilization Inca.

Pizarro said that I guessed this when I could not read the words engraved on the necklace. The professor puts on his glasses, closes his eyes to the pendant and says: "What are the words?"

Then he didn't need to see the words clearly. :

(You, miserable villain, who could have read a spell written not in an earthly language .. You, the man who awakened the curse of the Jackal ... You received the Jackal's revenge on yourself, and you cannot escape your inevitable fate .. The fire will devour your mortal body .. And then revenge will come true .. And from you there will be only a handful of ash from the wind.)

Then the professor's blood froze in his veins, and it seemed to him that he hears a rattling laugh ... Coming from the depths of hell.

Three days later, they found Professor de Toledo in a burning room .. Surprisingly, they found that his body had turned into a handful of Ashes, without affecting the rest of the contents of the room or even his clothes .. The fate line is another mystery of life gas.

There is a legend that belongs to the Inca civilization, a civilization that settled in Latin America for a long period of time until it was destroyed with the advent of the Spanish occupation of the continent.

In short, legend has it that the evil god (Jackal) had his wrath on the land of the Native Americans, the fire of his wrath on the greens and yabbah killed a lot of creation, the cattle was consumed, the plantations dried up and the earth was found (no doubt they were talking about an epidemic of epidemics that were important .. After that, the priests and the rest of the living prayed to the gods to protect them from the wrath of the Jackal and assign them to him .. The gods responded to them and managed to lock the Jackal in a statue with a pendant .. A spell was put on the necklace .. This is a spell , written in the language of the gods, can be read only by the unfortunate one who wrote "torment and anger of the gods" .. Chakala is freed from captivity and hunted down until he burns him with his fire.

A terrible myth, isn't it? .. But, despite the horror that gripped her, she remained, like any self-respecting legend, just a fantasy .. Until the real horror began .. And here the real beginning of our article can begin .. This dull, dull sentence appears (human self-immolation).
What is the definition of spontaneous combustion?
In short, a spontaneous fire in the human body without a known external cause, which can lead to several minor burns of the skin with smoke or the complete combustion of the body, only ash remains, but what is most surprising is the burning of the body and turning it into ash without the fire touching even his clothes or body.

Over 300 years, more than 250 cases of spontaneous combustion have been recorded, of course, it is not known exactly when this phenomenon began, but we are talking about the date of the first documented case of spontaneous combustion. Here are a few. :

1 - In 1637, a Frenchman (Jonas Dupont) published a book that included a study of this strange phenomenon, telling about an incident with Madame (Nicole Millet), a Parisian alcoholic, where she was found on a bed and left only with a skull, a finger bone and a handful of ash. Ironically, the sheet remained intact, and of course her husband was fingered before the court later acquitted him and linked the incident to an unknown person.

2 - in 1919, the body of a famous English writer (Temple Thompson) was found, the lower half of which was completely burnt without spoiling the clothes.
3. the incident with Mrs. Grace Pitt, an Englishwoman from Ipswich and also an alcoholic, where her daughter found her ashes without clothes.

4- In 1938 Mrs. Mary Carpenter caught fire within seconds while on a boat with her family and the efforts of her husband and children were unsuccessful in extinguishing the flames with sea water. Only her wet clothes were left without a trace of fire !!!.

On July 5-2, 1951, the remains of 67-year-old Mary Hardy Racer were found in her chair in her Florida apartment, where her skull and entire left leg were, and, as usual, a pile of ash,

which was about 5 kg and, as usual, also without fire affecting her clothes, the contents of the apartment or even a wooden chair and !!.
On May 6-18, 1958, the remains of Anna Martin's body were found in Pennsylvania, which consisted of part of her torso, shoes and a handful of ash, and her clothes remained intact, although forensic experts again reported temperatures between 1700 and 2000 m.

7-in 1966, the body of the doctor (John, 92-year-old Irving Bentley burned down in the bathroom of his apartment, only part of his leg remained separate from the body, and the shower curtains or paint on the walls did not suffer.

On August 8-26, 1974, baby Lisa's ashes were found in a hospital in Birmingham, England, and burned without any known cause or damage to her bed or clothing.
9.The most famous incident, witnessed by a large crowd of people, including the famous American physicist Hartwell, who stood there, unable to explain what was happening, caught fire in the legs and torso of a woman in Massachusetts, and then charred in front of everyone in a few seconds.

10-In 1938, when a girl (Phyllis Newcomb) was leaving the hotel with her friends, she was engulfed in sudden fire that hit her in a matter of seconds, and those present were unable to extinguish the fire that seemed to come from her very body! ! According to the investigator's papers, "I have never seen anything like it in my life: the girl was burned by a blue fire of unknown origin."

But wait, this brings up a very important question:

Did everyone burn out in the fire or survive?

And the answer is yes, there were survivors of this terrible fate, and here are two stories that can be told in the words of their heroes. :
1-September 1985 Debbie Clarke was walking down the hallway of her house when she felt a blue light emanating from her, which flashed periodically every time she took two steps. She found it funny. Her mother says: When I saw her, she started screaming furiously asking her to take off her shoes while her brother was jogging and we took her to the bathroom where they poured water on her. !!.

In the winter of 1990 in England, Susan Muched was standing in her kitchen in her fireproof pajamas when she suddenly caught fire and went out a few seconds before she had time to undress or call for help.

Jack Angel claims to have survived the self-immolation and survived, but without witnesses. Many people also claim that they woke up in the morning to find burn marks on their faces and stomachs for no apparent reason.

Until now, scientists have not been able to give a correct scientific explanation for this phenomenon, but they were unanimous that this phenomenon occurs with a person when he is mostly alone, although there are several attempts to give an acceptable explanation, but they are faced with the omission of some facts, and I I will list and refute them below. :

1- Religious interpretation: In the Middle Ages, the church claimed that self-immolation occurs with people who have reached the limit of their sins and deserve God's punishment by fire and destruction, while some priests argued that this is the result of meteors falling from the sky from the work of jinn and demons Although this interpretation was widely accepted among the people of that time, especially given the central role that the church played in medieval European societies, it did not explain to us the secret that fire does not affect anything other than a burning body, and did not explain to us why this is so. many people who have committed more terrible sins than just alcoholism have not done so. They took many lives in their villages without daring. ‘

2- Alcoholism: It is a common belief among people that the majority of burns were caused by excessive alcohol consumption, because alcohol is flammable. But recent experiments have shown that even bodies completely saturated with alcohol do not ignite in this way.

3-

4- 3. fat stored in the body: this belief was true because most cases were among obese people, where under certain circumstances fat stored in the whole body was burned, but dropped in thin people.

five-

6- 4. electrostatics: electric fields inside the human body that come into contact at a certain moment (short circuit) in an unknown way, as in nuclear reactions, produce enormous heat in a short time, but no form of static electricity is known yet. capable of leading to carbonization of bones, even if it is a lightning strike from the sky, and also not affecting the surrounding burn or even his clothes.

7- 5-hypothesis about the fusion of matter with its opposites: it is known that any physical particle opposes it and, for simplicity, there is also such a phenomenon as the fusion of positively charged protons (protons) with neutrons (neutrons) of equal charge and electrons (electrons) of negative charge in an atomic reaction. Some attribute this to the occurrence of chemical reactions of an explosive nature in the human digestive system as a result of malnutrition, which is the most acceptable explanation, despite its inadequacy, but explains three properties of this phenomenon:

8-

9- 1. Force 2. surprise 3. restraints

ten-

11- Finally, many writers were inspired by these mysterious incidents, the events of novels they created in an interesting literary style, for example:

12-1 -writer (Charles Brown) and his novel (Wieland) in 1978.2 is the most famous Russian writer (Nikolai Gogol) and his novel (Dead Souls). 3 - the most famous American writer (Charles Dickens) and his novel (Dark House) in 1852 4 - the most famous French writer (Jules Verne) and his novel (Captain Dick Sand) in 1878 5 - These incidents were addressed in one of episodes of the famous TV program (X-files) or (X-files).

thirteen-

14- In conclusion, I tortured my mind and searched for a long time for a convincing explanation for this phenomenon, but in the end I stood helpless, confessing despair in reaching a convincing explanation, and I appeal to all of you, if any of *you find a convincing explanation, provide it to all of us by quenching our thirst for knowledge and relieving us of the overwhelming anxiety that comes to us as a result of a terrible fate that one of us may suddenly face. And with all that I have made from excavating in dark caves of a phenomenon that no one I could not discover and reveal the truth and refute its darkness with the light of knowledge yet, but I came out with advice given to me by one of the sages after the search showed me and my efforts to trace the truth where it is. :*

fifteen-

16- (I advise you, son, not to say one day to your favorite in your will that NAR is my longing for you, my composition, who knows? Maybe your body suddenly caught fire ... And then you won't find someone, who will save you ... You will find Yulul next to your ashes only if they really love you ...)) !!!

17-

18- And how honest he was ... Who knows? ... May be....

Dream Series

The world of dreams .. For my age, it is stranger than the world of magic, and more mysterious .. Somewhere for a long time up to a third of my life in this world .. Our norm is to sleep from six to eight hours a day .. no doubt , this is for a long time .. And during sleep, gentlemen, something happens that we will talk about today .. Dreams .. And the world of dreams is a completely independent world .. Yes, it has much in common with the world of reality, but it is characterized by its own laws and symbols that make it autonomous .. A fact that requires its study and research .. The simplest thing that distinguishes the world of dreams and makes it radically different from the world of reality is that the laws of time and space do not apply to dreams .. Where all years and distances melt in the crucible of dreams .. In a clearer sense .. In a few minutes we can dream of events that spanned decades and took place in different places .. (Dreams mental health safety valve) is a scientific fact, confirmed psychiatrists and nev rologues .. Through dreams, the subconscious empties our repressed desires so that we do not strangle them .. The phenomenon of dreams has attracted the attention of both scientists and interpreters for centuries of history .. Although opinions on the interpretation of this phenomenon differ, it remains a difficult phenomenon to understand and interpret .. Modern science could neither penetrate into its essence, nor answer many questions regarding the formation and meaning of dreams ..

Historical overview

And review the history ... Some Eskimo tribes believed that the soul leaves the body during sleep and lives in its own world, and that the dreamer's awakening from sleep poses great danger, as it threatens to lose his soul and not be able to return to his body again .. In India, however, some ancient Indian tribes - who had the same faith - severely punished anyone who woke up in a dream .. In Pharaoh Egypt, the ancient Egyptians were the first to believe that dreams are a sacred revelation, and called them by the mysterious apostles to sleeping, to warn of punishment or consolation, consolation and foresight .. Some papyri have been found, including the papyrus of Chester Beatty - regarding its discoverer - from the Twelfth Dynasty with interpretations of dreams and their meanings .. As for the Greeks, Herodotus claims that in the country of the Greeks at one time there were about 600 temples dedicated to dreams and healing through them

How to dream? (Scientific review)

Dreams were subjected to scientific and laboratory studies to unravel the mystery of how and why they occur during sleep .. In some of these studies, patients and healthy sleep subjects were observed using electrodes that monitored the activity of the brain and nervous system during various phases of sleep. for a full night or more .. Dreaming has been shown to occur during a phase called rapid eye movement (REM), and this phase is characterized by increased heartbeat, palpitations, increased breathing rate of the sleeper and brain activity .. It lasts about 10 minutes and repeats every 90 minutes .. And, perhaps, this is what makes us dream, there are many dreams .. When we wake up, we remember only the last dreams .. During this phase, the eye drifts under the eyelid, and a so-called muscle relaxant appears .. Several studies from the University of Chicago show that dreams vary in length and can last up to an hour. and the old view of some psychologists that dreams are always short .. Dreams are also influenced by the environment and what we observe while awake .. University of California psychology professor Eric Schevitzgebel says that a survey conducted in the 1950s is the Golden Age black and white films - proved that most of those who lived during this period saw color dreams !! However, after this period and with the advent of color films, other survey participants said they were watching color dreams !! These dreams differ depending on the age, gender and nature of human life .. A student usually dreams that he is late for an exam ... Or he sits in an examination room and does not find a pen or a place for himself ... A person can have dreams that have something to do with external stimuli, such as a knock on the door, the sound of a phone call or a crying baby, but sees them a little differently .. For example, splashing water on a sleeping person's face makes him dream that he is standing under a hose or under a waterfall !! Scientists call this (interference of external influences in Dream Events)

Scientific explanation for sleep paralysis (jathom)

As mentioned earlier, the dreaming phase, or as it is called, the REM sleep phase, occurs during which the muscles relax .. For this it is worth mentioning a phenomenon that many may have experienced, namely (sleep paralysis) .. This is a truly terrible experience .. Some people wake up feeling the approach of death and the exit of the soul from the body, others believe that Genius is pressing on the chest .. Scientists explain that the mechanism of muscle relaxation ensures that you stay in bed during the sleep phase .. This mechanism ends as soon as you go into another phase of sleep, and sometimes the person wakes up during the phase of REM sleep, while the muscle relaxation mechanism has not stopped yet .. This leads to the fact that the person is fully aware of his environment .. But he cannot move at all .. Since the brain was in the sleep phase, this could lead to terrifying hallucinations, near death, or something else.

Daydreaming: ....
The state associated with dreams is called a waking dream, and through the label it is clear that this is a state similar to a dream, but does not arise during sleep ... A person is separated from reality into a dream state of imagination, imagining himself and having achieved many achievements ... Or he imagines who pretends to be a prestigious figure and lives in this atmosphere separate from reality and surrounding circumstances for various periods .. The problem here is that this state prevents a person from fulfilling his duties, where he spends his energy and time, and this state is often occurs in students preparing for an exam and in adolescents and young adults

Future dreams and the phenomenon of deja vu:
Surely this has happened to many before .. Experiencing a situation .. And then in a split second your body trembles with emotions and screams: you have already gone through the same situation before .. Exactly what I saw in my dreams .. And from us whose cry does not exceed its depth .. And it is we who confess it .. In any case, we have full confidence that our dreams will be fully realized in the future .. And make sure that we are worthy .. How to explain this phenomenon ?. And is it limited to some without others? ?.. Psychoanalysts call this phenomenon deja vu .. It is a literal translation of a French sentence (seen earlier) .. They see in déja vu an expression of a strong desire to repeat past experiences .. Doctors interpret this as a discrepancy in the brain that causes the brain to confuse feelings of the present with the past .. Scientists who study the supernatural believe that Deja Vu is related to past life experiences that we experienced before we came into the world in our present self .. This is another phenomenon known as BBE (pre-birth experience) .. Despite scientific advances at all medical levels, the phenomenon of déjà vu is still shrouded in mystery, despite the theories that have been attempted to explain it, none of which can be definitively defined .. Another worrying thing is that there is an opposite phenomenon of Des faux and as Jamais Vu (never seen before) feeling that is fashionably new and strange, as if he is facing him for the first time, and also this feeling does not last only a split second. It's like sitting with a friend or a family member, and for a moment it seems to you that you have never seen him as a stranger .. Developed several theories to explain the phenomenon of deja vu, the most important:

Dream theory: what we have been experiencing lately feels like a duplicate or another image from a past situation, it is just a copy of a dream that we already dreamed, but we have forgotten and do not remember .. We see dreams almost daily, but oh to human nature, we forget most of them and most of their details .. Therefore, some write their dreams in a go-go diary .. Some scientists also emphasize that sometimes during sleep a person goes through transparent stages of consciousness, soul and receiving Cosmic messages, that is he dreams of parts of his life or the future, as it happens in an honest vision .. Sleep is a partial paradox of the body ... Then the soul separates from the body and floats into space, meeting with other souls whose owners may be alive and possibly already dead long ago ..
Hence the logical explanation of the connection between living people and those who died long ago .. Through this comes another logical explanation of the phenomenon of evoking

the past and extrapolating the future .. These souls gather together in the so-called isthmus, and it is there that souls occur that we are not aware of with our visible mind .. We can not remember anything about him when we are awake, but it may happen that we accidentally hear some statement or see a scene that we think we have seen before .. More than one participant in the same scene or situations agree with the same vision .. Some scientists even say that before his creation, man was in a world called the world of atoms or the world of spirits .. In this world for him on Earth, many similar and identical events and events occurred .. So the events are identical in degree of correspondence .. Some believe that worldly life is only a version of this life, the second attempt given to a person to practice the concept of b lickiness to God and worship .. Therefore, when we go through an event or event and feel that we have experienced it before, this is a process that reminds us of the life that we lived in the world of atoms or in the world of spirits .. Some believe that a person is offered his life path before he is created in the life of God, and then breathed into his mother's belly, and then the scene of déjà vu can be repeated from the subconscious or from past memory in the first life .. This is consistent with the Islamic point of view. The Almighty says in his holy book: (Our Lord, our people are two, and we resurrected two, and we confessed our sins) this is also evidence of the truth of future visions .. Dreams bearing a prophecy about the future were mentioned in the Koran in the story of Yusuf (peace be upon him), so his vision when he was young was of what was later achieved with his parents and brothers, and then the king's dream of fat cows and skinny cows, which was interpreted as an early warning ... This was planned on the basis of this, before it literally materialized in the following years .. Also in the Qur'an there is a vision of Ibrahim (peace be upon him) killing his son Ismail (peace be upon him) .. But is everything seen before really seen in a dream ?. Actually not ... So there were other theories to explain the phenomenon of deja vu. I remember them:

1-The theory of memory centers in the brain: that there are areas in the brain and each area is responsible for the function, for example, vision is in the back of the head, and hearing is on the sides, and so on .. And what happens is that when you see something that is part of the visual center translates and its function is only to translate signals into an image and either understand and absorb this image, or remember it, if it is familiar, is located in another part, called the Cognitive Center, in this phenomenon sometimes there is a delay between two processes at a fraction of a second. Then the image comes to him later, and the brain thinks that he saw it before .. The nerves send these signals to the short memory center, where they are stored, and this process occurs very quickly (less than a millisecond) .. But sometimes it happens like this that the nerves mistakenly send the same signals to the center of long memory ... It seems to us that we have experienced this situation before .. For science, the center of long memory is a place where ancient events are preserved that can be recreated over time.

We all know that the brain has two or two lobes .. One of these parts is slightly in front or slightly protrudes from the other .. And when the brain receives any signal or image .. The two lobes take it together .. But sometimes .. This prominent or advanced part receives an image a fraction of a second before another .. Then it goes to another part, where all images, sounds, light signals, etc. are completely absorbed .. When a person understands a place or image in front of him ... He feels like he saw her before .. But the correct view is that she can be stored in short-term memory before being completely absorbed ..

3. paracological theory: scientists who interpret paranormal phenomena say that Deja vu is part of the sixth sense .. The human mind was able to predict an event in its image for a split second .. And when this event did happen, it seemed that the person had already passed through it used to be .. The subconscious has recently recorded information with the conscious mind .. Therefore, the person imagines that he has already gone through the same thing.

The role of dreams in inspiring scientists:
The dreaming function was not limited to reporting or warning the dreamer or his rebuke or other messages, but it alerted scientists to many ideas that aided discoveries and inspired inventions, such as:

Elias Howe, the inventor of the sewing machine in the eighteenth century, recounts: "That same night I had a dream in which people were throwing spears and each shaft opened into the outlines of the eyes above. This inspired me to find the right place for the hole in the sewing needle when I was designing sewing machine ..

James Watt, the discoverer of the so-called frame pelletizer used in mechanics that carries lead pellets inside, came up with the idea of this discovery after dreaming about it walking in a pouring rain of lead pellets. When he experimentally investigated this, he found that molten lead, if dropped from a great height, could actually turn into round solid objects or small balls.

....

Thus, we know the mechanism of the origin of dreams scientifically .. But what about the interpretation of dreams? .. Rather .. What is the purpose of dreams? ʿAnd what is its relevance? ʿRead the following .. People differ in their dreams .. The first group seeks to decipher their dreams and their connotations, indications and facts .. Another category avoids these dreams and prefers to forget them, or at least forget them .. But dreams remain in our reality and I do not show this because of the affection of many interpreters of dreams and Speakers of Visions who were endowed with this talent by God Almighty .. These interpreters were different from each other .. Some of them see that the interpretation of dreams as inspiration from God places it in some, but not in others .. This is a gift to which a person has nothing to do .. In turn, we find other people who firmly believe that the interpretation of dreams is rather a sharp mind and intellect than inspiration .. And the closest thing to say is, that the interpretation of Dreams is a sharp physiognomy and the ability to connect symbols and arrange events with extraordinary rapidity .. In addition to being an inspiration too .. Sometimes we do not see any connection between about brothers and events of a dream and its interpretation .. And we all know the story that happened to Ibn Serin, may Allah have mercy on him, when a kind man came to him and said to him: "I saw in a dream that I give permission." He said: "You are a thief, and Allah will return what you have stolen." He was asked about the difference between the two terms with this revelation, one said: the first is a good man, and he was the saying of God Almighty: (and the authorized Hajj came to people) and punk, and this was the saying of God Almighty: (then the authorized muezzin you are wrong, you rob people).)

Thus, the expression of visions from them is something inspired by God and thrown into the heart of the cross .. Some of them have a sharp mind and quick wit .. That is why we find non-Muslim visionaries such as Freud, Miller and many others .. You should also know that not every expression falls as it speaks of crossing .. He has already fallen, as it was said at the crossing .. And the crossing may be wrong .. We all know that Abu Bakr, may Allah be pleased with him, was great exponent, and yet the books of biography and history have preserved that he was mistaken in expressing some visions .. But there are those who are a little mistaken, most of his expressions fall, as he said, and this did not reconcile with him only a few human beings, such as the son of Sirin, may Allah have mercy on him ... Some of them also have little approval of the truth .. It is they who give a detailed expression to the events of the dream and sometimes even accurately determine the days and times .. And who became famous for this Abu Bakr, may he be pleased with him Allah ... may Allah bless him, he determined the events of the dream ... He is also expressive who sees that the dream interprets the phenomenon and is just brief .. There is no doubt that the first category is closer to the truth, especially if it is very honest and, as you know, little is wrong .. It remains to know that someone who has the gift of interpreting dreams never means that he is better or smarter than others, who does not have this talent .. In short, this is God's reconciliation .. For example, Ovais al- Karni, one of the followers, was better than the son of Sirin, may Allah have mercy on them in expressing visions .. However, it is not mentioned that he preferred Ibn Serin to Uwais or vice versa .. There are also stolen and said Ibn Musaib ... And they are all better .. This is the virtue of God, which He gives to whomever He wills, and God has a great honor ..

Types of visions: ....
Scientists have divided the visions into three parts:

First: true vision: true vision is a revelation and inspiration from God and has nothing to do with the fantasies of Satan or his will ... Not this confusion of the human mind, nor the consequences of the subconscious ... No events of the day or night about what a person cares about him in his vigilance and sees him in his pajamas - the so-called hadiths of the soul - and scholars differed ancient and modern in how to create a vision in a Man and is it in the mind or heart? It is true that the vision of the King creates in the human heart the gender beliefs of what is actually seen .. The mind translates them into symbolic signals .. This, in turn, is divided into two parts: 1-seeming sincere vision 2-symbolic sincere vision A seemingly sincere vision is one that does not need deciphering or analysis of dream events, but its interpretation is that they occur as they are .. Its sign is that it does not include symbols that usually observed in a dream .. Symbol-Faisal in the world of dreams .. There are hundreds of similar visions in historical books and speeches of predecessors .. Symbolic sincere vision is a vision that includes many symbols through their deciphering, analysis and connection with their interpretation. Expressiveness is not allowed just because it crosses the vision of a phenomenon so that skills and knowledge appear in its gut.

Second: the trumpet of dreams (Hadith of the soul): These are situations that occupy the mind and heart of a person .. These are also views that come from the subconscious and thoughts that a person holds .. Appear in the form of images and sounds, and the most difficult thing is to show the ability to distinguish the second the kind of true dreams are visions of sincere icons and unrealizable dreams .. They are very close .. But sincere vision has its connotations, and pressure has its connotations .. The first of its connotations is dreaming .. The second has psychological connotations .. In the sense that that through them the passage recognizes the personality, psyche and thoughts of the Dreamer, but the dream does not contain visions with intentional connotations, as opposed to the symbolic sincere vision

Type III: nightmare:

He is of the devil ... Satan manipulates a person, forcing him to portray terrifying scenes, images and sounds that have no meaning except intimidation .. At the level of psychiatry, doctors repeat similar nightmares daily to patients with anxiety, depression and schizophrenia ... The patient wakes up and is found in their worst psychological state .. This is especially true of schizophrenics .. Their nightmares are focused on hunting for ghosts ... It can happen in their vigilance .. And here is the difference between dreams and Nightmares .. Dreams are the release of internal inhibition and stimulation mind while sleeping ..

Now ... What do we do with our dreams ?.

A true vision of its two types should be praised by God and told only by a scientist or a person you trust, so as not to envy or harm you by misinterpretation .. But if you see a nightmare, do not tell anyone about it, and may Allah deliver you from it evil blowing to the left three times as soon as you wake up ..

Vision Interpretation Approaches: ....

When we look at the methods of interpreting visions or theories that relies on in hermeneutics, we find three prominent approaches in the world of hermeneutics:

1-metapsychological method: that is, a method that relies in the interpretation of dreams on things that are not related to traditional conventional psychology, and this method has two branches a: symbolic interpretation B: popular interpretation symbolic interpretation is one who looks and contemplates the apparent content dreams in general and is working to replace it with other content. An example of this approach is Sura Yusuf's interpretation of a vision of a king who saw seven fat cows eaten by seven skinny ones .. It was a symbolic alternative to the prophecy that foreshadowed seven years of famine in the land of Egypt, devouring the surplus of the seven years preceding it .. This dream was interpreted in accordance with the approach of the future symbolic interpretation, and scientists say: the success of interpretation by the symbolic method depends on sharp intelligence and discernment .. It is those who rely on this approach that say that the interpretation of dreams is based on physiognomy, not inspiration

A popular interpretation is called the thinking of the blade and its way in which it views the apocalypse as a fiction of writing code .. Every dream sign or intersection signal tries to transform into well-known signs and signals in accordance with a certain approach .. And who cared about this approach of the scientist (Artemidor). The success of this approach depends on the intersection and its ability to achieve the correct Approved Code .. Thus, there are many mistakes in this approach .. Therefore, the expression of visions using this type was not known to Muslims ..

2.psychological (psychological) approach):
This is the interpretation of dreams in accordance with the psychological climate .. The father of psychiatry (Freud) became famous for interpreting dreams in this way .. He gave us his book (Dreams), which says in the introduction: (reflections of human desires and fears in the subconscious are stable, while he is awake and his conscious mind watches them if he is asleep, and the forgetful conscious mind warns the subconscious mind expressing these desires and these fears in its own way, either explicitly, or by symbol and signal). He knows exactly what makes him uncomfortable and what makes him happy .. But this approach is not called what he makes expressive in the sense of expression .. It is the processing and interpretation of the so-called self-talk that comes from the subconscious .. Real ideas cannot be expressed with this approach

3-method of reasoning: that is, dreams arise for a reason and a cause and have a purpose and an effect, and this method, in turn, is divided into two parts a-suggestive reasoning: this means that the dream can be motivated by work and grandfather .. For example , the employee believes that his supervisor requires him to work hard until he gets a promotion, encouraging him to be vigilant until he finds and starts working hard. This type of dream depends on the idea that a dream is carries a message of vigilance to the Dreamer to take things and do things that are in his favor B-deductive reasoning: this type has two paths first: the path of the son of God's mercy values and the second: the path of the son of Sirin God's mercy path. The first example: clothing expresses religion and the vision of clothing and what was in it short or long, symbolized by religion, if the dress is long, then the religion of a person is strong and vice versa ... This is only an example and is not taken as a necessary rule for every vision in which dress or shirt second example: milk expresses a natural example third: a cow expresses good people for many a good cow and what is the use of it fourth example: a tree baptizes hypocrites

The path of Ibn Sirin, the Grace of God, is a path in which the transition is based on what is mentioned in the book, Sunnah and Arabic poetry .. For example, eggs in a dream go to women, because Allah says: (as if they are endowed with eggs) , and stones, because of the cruelty of the heart, say to God Almighty: (then your measured hearts are like stones) and other examples that we will show in the future, God will give ..

But before I continue ... I must emphasize and understand some things ..
1. The first thing I am aware of in this matter is what is circulating today in the hands of people of books that claim to include the interpretation of dreams, especially those attributed to the learned ancestors .. Like the book (Great interpretation) Ibn Sirin, this book lies to Ibn Sirin, may Allah have mercy on him ... He did not mention that he wrote a book about the interpretation .. And only in this case I say that Ibn Serin, may Allah have mercy on him, proved that he was the narrator of the hadith .. But he did not write the hadith of the Messenger of Allah, peace be upon him ... How can you expect him to neglect the writing of the hadith of the Messenger of Allah, may peace be upon him, and then devote himself to writing the interpretation of dreams ?! I think that this is probably the book his students have collected about him ... But he never planned it with his own hands .. And the trouble is in some books in which we read the explanation of everything in spelling !! Kktab (Anam aroma in the expression of a dream) Abdul Ghani Nabulsi 2-

visions do not have fixed rules, but differ depending on the seer, time and place .. Even for one person, the expression of vision changes depending on his situation and circumstances .. This confirms, that relying on these books and interpreting visions through them is unacceptable and incorrect .. But it is not allowed for scientists to mention, by the grace of God, that the expression of visions is like a fatwa, and it is not allowed to unanimously lie in a fatwa, he knew from this the sanctity of lies and charlatanism over people in interpretation visions .. 3. we must clearly distinguish between dreams or hadiths of the soul and True Visions, which came in maximizing more than thirty hadiths, including the saying: Peace be upon him: (there is no prophecy, but the evangelists said and what the evangelists said: a good vision sees Muslim or you see him) narrated by a Muslim The eye in the (continental mayor) speaks: (Prophets, peace be upon them, tell what will be through revelation and also a good sight the vision in which he tells the secret) 4.the vision, recently almost does not lie, as it is said in the hadith transmitted from Abu Hurayrah that the Messenger of Allah (peace and blessings of Allah be upon Him) said: "when the time approaches, the vision of the Muslim hardly a lie. "5. in direct visions are those with magic (in the last third of the night) .7-truthfulness of vision: Abu Bakr ibn Al-Arabi said that the mercy of God in his book (Everyday Al-Ahwazi): (Vision is the awareness created by God in the heart of a slave from the hands of the king or Satan) 8-visions with the interpretation of the Prophet (peace be upon him): a-interpretation of the Prophet (peace be upon him) milk in a dream by science B-wearing a shirt in a dream with religion B-running eye in a dream interpretation of sleep: 1-overflowing a bag of money conductor of death 2-bucket full of water symbol of a newborn 3-eat garlic conductor eat money Haram 4- Jasmine in a dream is a symbol of scientists and from the opinion of jasmine it is the death of scientists that captures 5-eat dates in a dream it is the sweetness of faith in the heart of a slave 6-dodgy interpreted as a journey 13-The apple of hope, work and leadership

And all my life .. Since God gave me the ability to interpret dreams and express visions, I have encountered many strange and miracles .. It is more surprising that I was offered many dreams, and their interpretation completely contradicted their content .. Until such the degree that the owner himself does not believe her until she already falls .. And then he will receive nothing but regret about what went wrong .. And dreams, gentlemen, signal the mercy of Almighty God .. Most of them- dreams of the future .. clearly predict what will happen in the coming days .. And in this they are either missionaries .. We announced a happy event as a relief for them and rejoiced after a difficult one .. Or be careful .. You warn us about the wrong that we do, and about cruel punishment for it by the grace of God .. Throughout my life I have met many applicants for Science and the ability to express Visions, who, unfortunately, are liars .. And also special problems that I often faced from -for mo her dream interpretation .. I explain dreams in detail .. Because of this, I have problems with the Dreamers, and perhaps the least of them is a mockery if I meet him in a tuxedo or do a bad deed or warn him about something- then .. And I, gentlemen, am not ashamed to admit it .. Many have already warned me about this .. Even close people advised me not to interpret such dreams that show the flaws of their owner and the temptation of his wrong actions .. But I took it upon myself the courage not to keep the knowledge given by God .. Therefore, I always thank the mentor and tell him: such dreams are a mercy for the owner to warn him before he falls into error or finishes what he does .. Here God's Grace is manifested through his servants .. So how does God want to warn the dreamer, but I refuse and leave him in darkness ?. Important .. In this article I decided to tell you about two miracles of interpretation: the first is the interpretation of certain things and things in a dream that do not really mean them ... So until the whole vision becomes clear to you ... And the second of them are some dreams with a strange interpretation, both those that I encountered in my life, and those that I read about in the books of the interpreters of my predecessor .. But this will be in the second part of this article, inshallah ... and even then .. I wish you a restful sleep .. And happy dreams ...

Oscar .. The cat that confused scientists

Cats are social pets that people have become accustomed to seeing around them since ancient times. Some love her and regret her beauty and intelligence, while others hate her for sneaking into the kitchen and spoiling his food .. There are also those who are afraid of her and do not feel comfortable in her presence, those who are especially afraid of her eyes , she sometimes looks so strange that it seems to her that she knows and understands, as if she wants to talk about something, about a secret carefully hidden behind those big shining eyes, about things and things that she does not know. Because of this mystery surrounding the cat, her name has been associated with many myths and legends since ancient times, they said that the cat has seven lives, they said that it can sense the presence of genies and ghosts, and that the black cat is a genie.

Of course, most people these days do not believe in these old superstitions and may have ridiculed them. But they may reconsider their calculations if they hear about Oscar, a tiny creature that has baffled scientists and made many wonder how far the line is between truth and myth.
An Oscar for the uninitiated is a beautiful cat whose strange story began in 2005 in the US state of Rhode Island. In the same year, one of the oldest and most prestigious medical institutions in the state, Stier geriatric care, decided to adopt a new cat to add to the collection of cats and pets that the sanatorium has long cherished and released into the corridors and halls of the sanatorium, in order to create an atmosphere of intimacy and fun that will benefit the psyche and health.

In order to get a new cat, some employees of the sanatorium visited a shelter for homeless animals, where they selected a small white-gray one-month-old cat, later named "Oscar".

Oscar lived on the third floor of the sanatorium, the floor dedicated to the care of patients with dementia, that is, this is the place where all hopeless cases, that is, elderly patients who are expected to die soon, fall.

Little Oscar quickly grew into a beautiful kitten in just a few months, and even though people have cared for and raised him since he grew up, he always tried to keep his distance from them, he was not that cat. who courted people and wiped his back with his feet in coquetry and dalal, and he would not sit in their house. No ... Oscar was not that kind of cat .. Rather, he was a mysterious cat, whose actions only a few of the nurses who helped him grow, preferred to be touched, sometimes disappearing for long hours, and then suddenly reappearing, wandering between corridors and rooms to check on patients, just like doctors and nurses do !.

As Oscar finished his first year at the sanatorium, the staff noticed strange behavior that suddenly changed and made him different from the other cats at the sanatorium. The face, as mentioned above, did not expect people and tried to stay away from them as far as possible, and, of course, would sit next to the patients in their beds, no matter what they tried to interfere with him. But his innate disgust for people did not extend to dying patients: wherever a dying patient was on the third floor, Oscar would suddenly appear, rise to this patient's bed and lie down next to him amidst general amazement, he sat quietly, resting his head on the patient's body, as if comforting and comforting him, and as soon as the spirit left the body of this patient, Oscar got up from his seat and immediately left the bed, as if the time between Oscar's appearance and the patient's death was often no more than two hours.

At first, none of the doctors and staff of the sanatorium believed that Oscar the cat could really sense and predict the death of people, which, of course, is impossible .. After all, he is just an unreasonable animal .. Right? .- No .. Ancient Pharaohs say no !! .. These great philosophers and spiritualists believed that cats are the living embodiment of the spirit of the gods; they believed that they see and hear what we do not see and do not hear .. That is why they respected and sanctified her to such an extent that they mummified and mourned her death.

Of course, the doctors at the staircase sanatorium would not have believed what the pharaohs said. The only logical explanation that seemed convincing to them is that the increasing appearance of Oscar with the death of patients is just a coincidence .. No more. But Oscar's coincidences became more and more frequent every day, which further puzzled and overwhelmed everyone ..

Once in the sanatorium there was an Elderly Lady with caked blood in her leg, she was in a coma, and her leg was stiff and cold due to the lack of access to her blood, but the doctors did not expect that she would die soon, or so they thought, until Oscar did not suddenly appear and did not go to bed, did not sit at her wounded leg and did not surround her. The old man was dead! ..

Oscar entered the patient's room again, went to bed and quietly sat down next to the old patient who was sleeping, but the patient's family did not like Oscar's presence, they were afraid that his comfort would be disturbed, so they asked to be taken out of the room, and indeed, alone of the nurses held him gently and then threw him out of the room. But what happened next stunned everyone: the stubborn kitten did not leave, he began to moan with a screaming sound, he knocked on the door and scratched it with his claws, as if asking permission to enter, and as soon as they opened the door again, he ran and jumped back to the bed of a sick old man who died within the next two hours. This time, the doctors did not expect the patient to die either.

The controversy over Oscar's abilities at the sanatorium grew day by day. The team they believed in and the team they categorically rejected. In the end, both teams decided to put Oscar to a test that would surely remove their doubts. In one of the rooms on the third floor, an elderly patient was dying and he was supposed to die in less than an hour, so they brought Oscar to this patient's room and sat him on the bed. They said that being near the dying patient would certainly confirm him supernatural abilities. if the opposite happens, that is, if Oscar gets out of bed and leaves before the patient dies, it will mean that he is not able to predict death and that all this is nothing more than a coincidence.

Against the backdrop of joy, skeptics and believers see Oscar as he got out of bed after several moments of putting it on, which is counted by a team of skeptics as irrefutable proof that Oscar's talent is allegedly not healthy, and not at all because he could not recognize this patient in the case recent conflict.

Contrary to all expectations, the dying old man did not die within an hour, as the doctors expected, but surprised everyone and clung to life for another ten hours. Shortly before his death, Oscar reappeared, this time on his own, accepted the old man's bed gracefully and sat down next to him calmly, as he does every time, after less than two hours of the old life's difference, which amazed both supporters and skeptics. Oscar surpassed doctors and their sophisticated equipment in predicting the exact date of a patient's death.

Over the next two years, Oscar predicted the death of 25 patients. No one doubted his ability, but the staff of the sanatorium watched his inspection trips with great interest, and the fact that he was sitting next to one of the patients was a sign that this patient would soon be surrounded by death. The hospital staff quickly called the patient's family, asking them to come immediately because their patient would die within hours. Despite Oscar's fame at the sanatorium, the world did not hear of his prodigious talent and ability until 2007, when Dr. David Dosa wrote an extensive article about him in the Medical Journal (NEJM), one of the oldest and most sober medical journals in the United States. Since then, Oscar has become the center of attention of the press and people.

The sanatorium staff thought that Oscar's fame and notoriety would make the patient families angry and disapprove of his presence around their patients, but on the contrary, everyone noticed that these grieving families were grateful for the presence of a beautiful cat with their loved ones in their last hours. seeing him as an angel of Mercy, whose presence would ease the pain of dying, and, according to the Stear Sanatorium website, Oscar predicted 50 more deaths before 2010.

Scientific explanations
Of course, scientists would never believe in superstition and would never accept the explanation that cats have a superpower to sense and determine the movement of

invisible etheric energies and auras, that is, what people identify with the soul, Jinn and ghosts. On the other hand, scientists believe that animals already have superpowers that humans do not have, not metaphysical or metaphysical abilities, but sensory-sensory abilities. No one disputes that animal senses are many times stronger than our senses, especially smell and hearing, so the most rational interpretation of Oscar's supernatural abilities focuses on the sense of smell of cats. Dying cells begin to release certain enzymes and chemicals into the human body shortly before death to prepare the body for the decomposition phase that follows death, and scientists believe that Oscar, through his presence in the sanatorium, was able to develop a special talent for distinguishing the smell of these enzymes and chemicals released by the body at the time of death, and by capturing this smell, it can tell and determine when patients die.

"Maybe he can distinguish a specific smell associated with death," says Dr. Dosa, commenting on Oscar's abilities. "I think there are certain chemicals that are released into the body when you die, and Oscar can smell and distinguish the smell of these substances," says Professor Joan Teno, a university professor who specializes in health.

"I think it smells like certain chemicals that are released into the body before dying, cats can smell a lot that we humans can't smell, they can even smell disease," says Margie Skrek, animal behavior scientist from the University of British Columbia.

Another view of Oscar comes from another zoologist, Dr. Daniel Estep, who says: "One of the characteristics of people who are about to die is that they don't move much, and Oscar may have caught on to the fact that people who are dying are very quiet. and quiet. I think it's not the smell or sounds of the patient, but the movement. "

Speaking to the BBC about the ability of cats, Dr. Thomas Graves, professor at the University of Illinois at Urbana-Champaign, said: "Cats often feel sick from their owners and even from other animals. He can feel the changing atmosphere and is also known for his great ability predict earthquakes. "
Sensitive creatures .. But ..
In addition to the opinions of scientists and doctors about the well-being of cats, cat breeders around the world believe that their cats are the most unique of all animals in terms of sensation, emotion and behavior. Unfortunately, this does not explain Oscar's extraordinary talent and abilities .. Why? ..

Just because Oskar does not live alone on the third floor of the Stir sanatorium, there are six more cats on that floor, so if it's about smell, why don't the other cats show any similar abilities to Oscar?

And even if we claim that it is the smell that makes the kitten come to the beds of dying patients and sit next to them .. Does he see what we humans do not see? ... Does he understand the idea of death ?! Does he entertain and help dying patients?

Remember, dear reader, that we are talking here about patients with dementia, that is, they are not aware of anything that is happening around them .. Personally, I do not *understand this and I can not digest the hypothesis that this stupid animal really understands what a human death.*

Dream Series - Miracles of Interpretation
        1.   *vision of the dreamer's death:*
** Anyone who dreamed that he died without a body from death or illness lengthens his life .. But his house is crumbling !! .. If he sees that he is buried without tears and burial, he will not be able to rebuild his house .. He will leave it to build .. And those who saw him as if he had not died were condemned !! * The one who saw that he was dead on the carpet that was extended to him .. If he lies dead on the bed, they will give him a lift and give him*

*a place, and he will receive good from his family !! .. There are not enough people who saw how he died naked on the ground .. * If a pregnant woman sees that she is dead and is worn around her necks, and people cry for her without a voice and without ways, she gives birth to a soft birth, and a baby boy, and she is happy with it !! Anyone who sees that he is dead and remains single will marry .. Anyone who sees that he is dead and married will divorce his wife .. In addition, the fact that one of the spouses saw the death of the other indicates that that there was a divorce between them. * The one who saw that he died and then lived again in a dream commits a great sin and must hasten to repent for his words: "Our Lord, our people have two, and we raised two, and we confessed our sins." he is content and content in this world ..*

2. The Dreamer's vision of the death of another:

** Saw death is the guide of a destroyed country and the present and vice versa ... * Anyone who has seen a dead person knows this as if he died again and had ablution, funeral and commemoration, this is the death of one of the people of this dead person. But if they cry for him without a voice and without a path, he will marry the people of that dead woman and live happily with her !! * Whoever saw that he was found dead, he finds money that was hopeless * He who saw that his son died, got rid of his enemy !! .. For those who saw that his daughter died, despair of the vulva !! .. If he sees that he is the creator of the dead, or touches them, then these hypocrites hurt him .. * Whoever saw that he ate dead, he lengthens his life !! .. And if he carried the dead on neck, he received a lot of goodness and money !! .. * If he sees a person alive, as if he died in a dream, then this person is a proof of God and the difficulties of his life .. * If he sees that someone is telling him, that Lana is dead, and this person is actually alive, then this person becomes sick and can die in him ..*

3-See the Dead Alive .:

** If he sees that his dead grandparents have returned to life or one of them, he becomes hardened and hardened in his work and life, despite his difficulties ... * If he saw that his dead mother came back to life, vulva came to him who he was .. And if he sees his dead father, he comes back to life .. Only the vision of his father is stronger .. If he sees that his dead son has returned to life, the enemy will appear to him from where he is not in score!! And if he sees his dead daughter come back to life, the vulva will come to him !! * The one who saw his dead brother come back to life is strengthened after weakness to speak :) I emphasize this Azri (* the one who saw his dead sister came back to life, this is the arrival of his lack of travel and pleasure comes to him behind the saying :) and she told her sister a story so that I could see him from the side) * if he saw his dead uncle or aunt. * If he sees that he is alive and dead, he is guided by an unbelieving or disobedient God .. And if he sees a dead naked, then this is his exit from the world without guilt, just as if he saw a skeleton .. My dead nakedness is proof of his comfort .. In addition, if he sees a dead man sleeping, he feels comfortable in the end .. If he is seen alive and dead, he will betray an unbeliever or repent of lust .. * If he sees a dead man wearing green clothes or a crown , he will receive a certificate of his death ... * If he sees a dead person laughing and then crying, or sees his blackened face, this indicates his death as blasphemy for his saying: (As for those whose faces are blackened, you do not believe after his faith) and vice versa, if he sees the dead laughing and bright in clean clothes .. This will also be evidence of the wealth of the people of the dead after their poverty .. If the deceased is one of his own* people, he will cleanse him after his death and reach his pr to this deceased ... * And if he sees the righteous dead, as if he was frowning at him, then he can reconsider himself and warn, then he will commit a sin .. * If the deceased sees dirty clothes on him, it means that he has there is a duty that he has not fulfilled, and he is responsible .. If he has no religion, and he is good, then his people followed him and commit atrocities .. * If he sees that the dead is busy or tired, he fills it with his score .. If you are good, then it is filled with his family and relatives who are still alive .. * If he sees a dead man, he knows that he died blasphemously and saw him at his best, that is, evidence of the Islam of his family, and not his good hearts ... * If he saw the dead, he was rich in the world, and saw him poor, then it was his lack of favor ... and vice versa ... * If he sees a dead man praying outside his prayer place in this world, it is constant mercy for his soul .. If he sees him praying in the same place as his prayer in this world, then his people will adhere to their religion .. If he sees him in the mosque, it will save him from torment .. * If he sees a dead man complaining about his head, he asks about his failure in the case of his parents or his boss .. If he complains about the right of his wife, if he complains about his belly, he asks about the rights of his father and relatives and about his money, if he sees that his man is complaining, he asks about spending his money without God's consent, if he sees that he is complaining about his hip, he asks about cutting off his womb, if he sees that he is complaining about his legs, he asks about his life in a lie ..

3. Follow the dead and conduct a dialogue with him .:
* If he is seen as if he follows the dead and follows his trail in his entry and exit, he will imitate his actions of righteousness and corruption. * But he felt how the dead called him from where he did not see the answer, and went out with him so that he could not refrain from dying in a disease similar to the one that the dead called him or in his death, destroyed or drowned or suddenly .. And if he sees as if he followed the dead, he enters with him into an unknown house, and then does not leave it, he dies * and if he sees as if the dead tell him that you are dying in time of this, he says it correctly * and if he sees as if he followed the dead and did not enter the house with him, or he praises his vision, and if he sees as if he is talking to the dead, he lives a long time and works for people after a dispute, and if he sees as if he is accepting an unknown dead, he gets money where they do not count .. The pre-dead knew this from the dead with his knowledge or his money the next dead and his family are good * but felt that the dead preach or teach him a note about it affects his kindness in his about religion as much as this

4. Atiyah al-dead for the living and Atiyah al-Hai for the dead:
If he sees that the dead have given him something of worldly pleasure, then he is better .. Ibn Serin said: "I like to take from the dead and I hate to give it" and said: "If the Dead takes something from you, that dies), therefore, everything that the neighbor sees that he gave to the dead, he does not like, except for two things: firstly, if he sees as if he gave the dead a watermelon, it means that he gave the dead a watermelon.

If he sees as if the dead gave him a new shirt or a clean one, he gets a good life, like the living dead in the days of his life, and if he sees as if he gave him a crown, it hits Jahu like Jahu. If he gives him a patchwork dress, he lacks, if he gives him a dirty dress, he drives obscenities, if he gives him food, it hits halal, whence he doesn't count, if he gives him honey, he gets loot, whence he doesn't count.

* If he sees as if he gave the dead clothes that he did not publish or wear, it is damage to his money or illness, but he heals .. If he felt like a dead livery and gave her to the dead, put her dead, she dies the hand did not tell her or give her did not hurt him * but felt that the dead bought food, that the food was boiling or wasted * but felt that the dead were selling food or movable property, food and a maze

* Finally, the deceased saw him doing a good deed in a dream, for example, praying or reading the Koran or alms, etc. The deceased reminds him of this work or encourages him to do it, especially if the deceased is known or from his family, because relatives will be the first to know about it ..

5. Nikah is dead:
You may be horrified to dream that you are fucking and sleeping .. You wake up dazed .. They are still being reprimanded .. But what does that mean? ... Here's what we're about to find out ..

If he sees the sleeping man, as if he should fuck an unknown dead man in the grave, he will betray him, and if he sees that he should fuck a man or a woman, he will find a need from which he can despair. If he sees that he should fuck his dead friend, he would like the people of the dead to help them. If the dead is an enemy, he will reconcile with the people of this dead man .. * If he saw that they deny the holiness of the deceased, if the Dreamer really does it, that the deceased honestly continues or prays. Or his family comes, wants them and spends on them ... If the Dreamer is disobedient to God, then they say that he is offered Haram .. And also if he sees as if a dead woman, as you know, fucked him for his work or of his money * if he sees as if a dead woman is alive, fucked her and hit him out of her water, he earns his need and spends money with the same kindness from him and gets a renewed mandate and a profitable trade * if he is. And if he sees as if he had married a dead woman, and he sees that she is alive, and entered her, and did not touch her, but turned to her house and reassured her, the visions showed his death .. In all this the vision of a woman follows the path of the vision of a man ..

6.crying and whining:
Contrary to all expectations, crying in a dream is considered a desirable thing if it is not accompanied by barking and screaming .. Whining is an omen and a bad omen .. And he (Sirin's son) said: "Crying in a dream is the Root of the eye ... But if crying is combined with a groan or a dance, then this is not praised .. If he sees that the ruler is dead, and people bark at him and throw dust on their heads, then this ruler is unjust and unjust .. * If he sees that the ruler is dead and people cry for his funeral without shouting, then he is a just ruler who pleases people with his rule ..

7.washing dead:
Since washing is usually cleared of dirt and it occurs in a dream from a carefree and elevated business ... * The one who saw the dead, wash or wash himself, this will be the pinnacle of the care of his people and their wealth and their way out of worries .. who sees himself dead and washed, will be handed over to an unbeliever or a repentant punk ... * The one who saw him washing the clothes of a dead man is famous, he does a running charity for the spirit of this dead man ..

8-casing and:

Unlike ablution, the shroud is an invitation to adultery, debauchery, sin, if it does not cover the whole body .. Anyone who sees that he is called to wear a shroud, he is called to adultery, and he will not respond with gratitude to God .. What is shorter than shroud, the closer the Dreamer is to repentance and vice versa ..

*But if the shroud covers the whole body, like the body of the deceased, this indicates death ... * The one who saw him wrapped in a shroud, as the dead were wrapped, his visions indicated his death ... * And the one who saw that unknown people decorated him and dressed him in fancy clothes for no apparent reason, such as a holiday or a wedding, and that they left him alone in the house is proof of his death.*

** Regarding embalming (from musk and other things), a vision in a dream is a good guide, because the purpose of embalming is to prevent the corruption of the dead*

9-coffin and funeral:

*As usual, the vision of the coffin is contrary to expectations .. So the origin of the coffin of recovery !! .. * The one who saw that he carried his great order and most of his money on the coffin and received a rule that conquers the necks .. Even his followers or the servants became like the number of people he saw spreading it !!*

*The funeral has two faces: Mahmud and the disgraced .. Al-Mahmud is: * to see how the person himself is carried on the coffin, as I said, and the mourners cry or remind him of him and pray for him ... He will rise above himself and will rise above his place in the world .. * Or to see as if the funeral arrived at the cemeteries, this is a right that reached his parents after they took offense at him ... * The one who saw as if the coffin walked through the air , it was the death of a scientist or a good person, whom people forgot about him and did not appreciate ... * The one who sees that he is carrying a dead person and brings him to the cemetery, let him know that he is right in his order ... Let him continue and doesn't care about opponents ..*

*The rest of the funeral vision in a dream is corruption in religion and hypocrisy: * who saw his funeral and mourners revile him, this is the corruption of his religion and a consequence of his order in the world .. He is a harbinger of God to warn and reconsider himself ... whoever sees that he follows the funeral follows the corrupt ruler .. * The one who saw the funeral somewhere is the hypocrisy of the people of this place ... * The one who has seen a place with many funerals, the people of this place often perform atrocities .. ** The one who saw how he carried a dead man and delivered him to the market is a merchant whose product will bring great profit .. But all his money is haram ..*

10-prayer for the departed .:

*He also has two opposite sides: * the one who sees that he is praying for the dead Mom is more than praying for this dead and asking for forgiveness, and his work reaches this dead .. * If he prays for the dead, then he imam who received a position and mandate from a punk ruler !!*

11-burial:

He has many different facets: it either means a long journey ... or prison ... or death ... or get rich ... or a new home ... and even marriage !! .. The meaning even varies from whether the Dreamer sees himself buried or sees another ... and it all depends on the specifics of the dream itself ... Here are the details .:

The Dreamer's Vision to pay: ....

** Anyone who sees that he is buried without death and with depression, imprisoned .. If he ever saw himself died in his grave, he died in this prison .. If he did not die in this grave, he survived and was released from prison .. In general, a dug grave is a prison dug for you or for others .. * The one who sees that he died and is buried, he travels far, infecting Mala !! .. The one who saw that he was buried (put in the grave), he is glad that God gave him a new home !! .. So the people of the Goldsmiths put dirt on him, God gave him as much money as this dirt !! .. * He who saw that he was buried in a large grave in white and good clothes .. Then his parents visit him in the grave. he is getting married !!*

B-vision of the burial of others: ....

** Anyone who believes that he buried another or buried him, will leave the buried one to perish because of him ..*

12-graves:
** A-digging a grave: ....*

*Whoever sees that he is digging his own grave will build a new house .. * If he sees a grave on his feet without a funeral, he will buy a new house ... * If he sees that he is digging his own grave on the surface, he will live long. . * If he sees that a famous dead man has turned his grave into a house, then the people of the Dead will build a new house ..*

B-visit to the graves: ....

*As I said, an excavated grave is a prison .. Therefore, whoever sees him visiting the cemeteries, his relative or friend will be imprisoned and visit him in his prison !! If he sees himself wandering among the graves in their known place, he does not notice the truth and follows the lies .. * If he wanders between the graves or a grave in an unknown place that he does not know, he mixes with hypocrites, beware of them ... * If he sees himself or others wandering among the graves and surrendering to their families, he will be a beggar and beggar, if he or he sees him ... * Anyone who saw as if he was standing on the grave, he commits a sin, may Allah fear and let it be so as to say: (and do not stand on his grave.)*

B-exhumation of graves: ....

** The exhumation of the graves is the revival of the year of the burial of the deceased in his grave .. If the Dreamer sees him digging the grave of an infidel, he will follow his year and get lost .. If he digs the grave of a scientist, he will follow his year and be guided .. If he will dig up an unknown grave and lead to a worn-out corpse, he will be lost.*

Exhumation of the Tomb of the Messenger of Allah (peace be upon him) is the revival of his year and steadfastness in religion ... Especially if the Dreamer is the Messenger of Allah, peace be upon him alive in his grave ... But ... If the Dreamer digs the Tomb of the Messenger of Allah, peace be upon him, then he finds a corpse or bones that are worn out and may Allah be grateful .. He is a heresy, let him reconsider himself and let ..

13.the vision of the king of death, peace and blessings of Allah be upon Him:
** Who saw the king of death a long time ago !! .. If he saw him satisfied, he would have concluded with joy and God's satisfaction, and he could have died as a martyr or receive a degree of martyrdom .. But if he sees him dissatisfied, he warns of bad imprisonment and punishment by God ...*

... .

And now it's time to be practical !! .. Now I will show you some visions and dreams that I have interpreted and expressed in the past to their owners .. I will show you as written by the owners, but without expressing it, waiting for you to express it in your answers .. Consider it a competition or a challenge .. And let each of you feel this ability in yourself .. And judge yourself .. And the question is, can I interpret my dreams? ..

... .

1-I would like to interpret a dream that I see a dead person, whom I know, who is really dead, and I see him dead in a dream, but as if he is alive and very sick ... .
... .

2-Peace, mercy and blessing of God I dreamed that I was hit by a car and I died, and they all knew that I could die, I remember that at school everyone was just sad for me and my friends, I bet they were crying there for me .. Then my mother found out that I was still alive .. When I entered the school, I found that everyone was afraid of me, because they heard that I had died ... Oddly enough, I told them: No, I died ... You see my ghost, and the dream is over .. So, in the dream you are alive and dead at the same time .. Please explain, God bless you.

... .

3.my sister dreamed that she would write her book about a deceased person, and the author who will write this book also died ....
....

And after ... Now that I've shown you what I suggested, let me ask you: Will we continue this series of articles together or move on to another topic? Together we strive to reach the point where you can express your visions and interpret your dreams for yourself .. Because I will not be able to fall into the circle of hypocrisy, break my promise and interpret your dreams as I said .. I do not hide the fact that now I am writing my memoirs, which many of you have asked me to write and submit .. If you want us to continue this series, I will postpone the publication of my memoirs until a later date .. If you want otherwise, you are welcome .. It is your decision. I'm waiting for your opinion and .. You're all right.

The mystery of the disappearance of the Neanderthals

Life in the Stone Age was not very good .. I say this to those who remind us of the old days and just puzzle over the meekness and splendor of the past .. Both, dear reader ... Were not at all pretty .. It was frantic, constant chase interrupted by short periods of rest .. You had to run all the time .. Run .. Run .. You have prey .. Or behind you are ravenous fangs seeking to despise you. And in those few moments when you stop running, you have to hide deep in a lonely cave .. And cool .. And dark .. Where should you eat, sleep, defecate and fuck your partner in the same small place in front of of his family gathered around this fucking pitiful hearth .. What a life! .. And you are very lucky to survive until you are thirty .. Life was short .. Human numbers are small .. Thank God we were not created in those dark ages that I briefly visited in this article .. So what about to come with me, dear reader? .- Do not be afraid .. We will not stay long.

Human history is permeated with riddles and riddles, this Reasonable creature, called man, lived and reigned on this earth from the moment of its appearance more than two hundred thousand years ago, this is what modern science says. Unfortunately, we know almost nothing about what this man did and did during those long and prolonged periods, since our vast information about him is limited to the last five thousand years, that is, from the moment of the emergence and birth of the first human civilizations on the banks of the Nile , Tigris and Euphrates, as well as the accompanying invention of writing and the beginning of written documentation of events, customs and beliefs in force at that time. Before that, before the invention of writing, our knowledge was almost limited to ancient stone tools, pottery, and skeletons scattered here and there in caves, cemeteries and caves. It is for this reason that the eras preceding biblical documentation are considered one of the most mysterious, dark and controversial periods in human history. Not surprisingly, Neanderthals are one of the most enigmatic mysteries of prehistoric times. Since the discovery of his first skull in Bulgaria in 1829, opinions, theories and hypotheses about him have ranged to contradiction and disagreement, leading to many myths and mistakes about him.

Perhaps the most common of these errors is the belief that Neanderthals are the ancestors of modern humans, and that our cave-time ancestors were Neanderthals. And for my age, this is one of the biggest and biggest mistakes. We and Neanderthals are two different types of people. It is true that we belong to the same genus, called Homo, that is, man, and it is true that we are all cousins descended from the same ancestor who lived five hundred thousand years ago, that we share 99.5% of our genes, and that some of us still have a small amount of Neanderthal genes in our genes ... But beyond all these similarities, we imagine humans in a completely different species known as Homo sapiens and Homo neanderthalensis.

Another common mistake is the belief that Neanderthals were a species before modern humans and that they did not live together in the same place and did not have any rapprochement with each other, which is also not true, the two species have lived side by side since the arrival of modern humans on the European continent near 43 thousand years ago before the disappearance of the Neanderthals.

Neanderthals were neither monster nor ape.
Recent discoveries and technologies have refuted many of the previous ideas and conclusions of scientists about Neanderthals and began to paint us a completely different picture than the one that hung in our minds about him. The process of repainting and shaping his skull using a computer proved that his features were closer to those of a modern man than a monkey, with a slight protrusion above the eye and a slight indentation in his forehead and chin, and he was not a thick-haired creature, as some are trying. depict, recent studies show that his hair was thicker than that of the monkey.

Well, if modern man and Neanderthals are so similar .. So what is the difference between them and why do scientists classify them as two different ?.
In fact, there were several obvious differences between the two. Recent discoveries, especially the rare skeleton of a Neanderthal infant found in the Dodriya cave near Aleppo in Syria, has shown that the brain of a Neanderthal was the same size as that of a modern human at birth, but that with age, the brain of a Neanderthal was larger than that of a modern human. But this superiority in size does not necessarily mean that Neanderthals were smarter than modern humans, scientists say that the way the brain grows and develops after birth is what plays a key role in determining the amount of intelligence, and the way the brain develops, according to scientists. was poured in favor of modern man at the expense of the Neanderthals, but the latter also had evidence that he was burying his own death. ‹What he first painted on the walls of ancient caves, the oldest human drawing, the one on the wall of one of the caves in Spain, was probably made by Neanderthals who have inhabited these places for millennia.
Structurally and physically, both species were similar in stature, but Neanderthals were much stronger than physically modern humans, their hands were powerful, as if they were an Adagama with his broad chest. On the contrary, modern man surpassed the Neanderthals in terms of agility and speed, the Neanderthals walked hard and slowly due to the nature of their bone formation and, perhaps, seemed to the beholder as if staggering from side to side when walking and needed more energy when running, jumping and chasing , so modern man had the advantage in any possible confrontation with the Neanderthals.
Neanderthals also differed in their hair color, some scientists say, who found a gene called (MC1R) in a DNA sample taken from Neanderthal bones, and according to the researchers, this gene gives hair a characteristic red pigment, and red hair is known to be rare color in modern humans, so Neanderthals were very rare.
In addition, Scientists believe that modern man also surpassed them socially, living in larger and more organized groups in terms of the division of tasks between men and women, and surpassed them in terms of language development and communication with their own species, although modern scientific opinion suggests that Neanderthals may also have been able to speak and have a linguistic understanding. This point of view was supported by the discovery by scientists in 1983 of the hyoid bone inside the remains of the bones of the Neanderthals, this horseshoe bone is located in the front of the neck below the chin and its presence is necessary for speech and speech. What's more, scientists recently discovered the FOXP 2 gene in a DNA sample taken from the bones of Neanderthals in northern Spain.

The bottom line is that if we lived in dark and cold prehistoric times, we would distinguish between Neanderthals as easily as our ancestors, due to obvious differences in general appearance and behavior.

Why did the Neanderthals die out?
Since its discovery in the 19th century, scientists have questioned the mystery and cause of the Neanderthals' extinction some 24,000 years ago, leaving room entirely for modern humans to assert their control - and evil - over this vibrant planet. To answer this question, scientists have several theories and hypotheses that we have sold in the next few lines.

Conflict theory with modern man
Neanderthals were physically adapted to live in cold regions, so their range was limited to a geographic area stretching from northern Iran, Iraq and the Levant through Central Asia, Russia and Europe to the Mediterranean coast of Spain, where caves in the Strait of Gibraltar were his last refuge. A Neanderthal dwelling in such vast expanses and its owners entertained for hundreds of thousands of years, until the dehydrogenases of his dominion of another type of people suddenly appeared on the stage behind a rival place in everything. This second species was, of course, none other than modern man, who first crossed Africa to Asia and then headed to Europe about 43,000 years ago.

According to some scholars, modern humans, who were more organized and lived in large groups, gradually exterminated and expelled the Neanderthals and drove them to the West and South of Europe, where they found their last refuge. This view among scholars is reinforced by the often violent, violent and violent behavior of people over the centuries in dealing with rivals, the most obvious example being how winners dealt with losers in wars and conflicts, ancient and modern, and as whole indigenous peoples and tribes. America, Australia and Africa have been exterminated over the centuries. In addition, hundreds of animal species and species have been exterminated over the course of human history with the smell of blood and death.
The theory of marriage with modern man
Contrary to the previous theory, this theory states that Neanderthals became extinct due to their mixing and mating with modern humans, that is, their offspring were lost and scattered within the offspring of modern humans, a well-known situation that has occurred repeatedly over the centuries, small and dominant groups tend to dissolve and merge into larger groups, and history has seen many tribes and peoples disappear.
In 1998, scientists found a skeleton of a child in a cave near the Portuguese city of Lisbon, which they named the child Lagar Velho , and this discovery was significant because the skeleton of a child had mixed features between modern humans and Neanderthals, probably a hybrid.

Most European and Asian peoples, as well as light-skinned North Africans, carry 1-4% of Neanderthal genes, which are not carried by the African race because Neanderthals have never lived in Africa. This small fraction of genes indicates limited marriages between modern humans and Neanderthals during the Paleolithic era.
Climate change theory
According to this theory, climate change during the last ice age more than fifty thousand years ago played a major role in the extinction of the Neanderthals, during this brutal era,

turned much of Europe into polar deserts, and this led to the decline of forests and green spaces that were grazing animals. , which the Neanderthals hunted, and with the decrease in the number of these animals, the survivors gradually retreat to the warmer shores of southern Europe, especially to the Iberian Peninsula, that is, in Spain and Portugal, which are known as the last refuge of the Neanderthals.

Climate change was not limited to the onslaught of cold and frost, but was exacerbated by the eruption of several European volcanoes during the same period, and these volcanic eruptions, with their black clouds and volcanic dust, led to the destruction of most life forms within the Neanderthal existence. Thus, the union of bad weather with volcanic eruptions weakened the existence of Neanderthals and severely depleted their numbers, and therefore modern man faced neither competition nor resistance from emaciated Neanderthals when they first set foot on the European continent more than forty thousand years ago.

Theory of diseases and epidemics
As noted earlier, Neanderthals and modern humans are very similar, which is why Neanderthals, like modern humans, were prone to deadly diseases, diseases and epidemics. Some scientists argue that its lack of natural immunity to certain types of microbes and viruses that modern humans brought with them to Europe made it an easy target for the many deadly epidemics that weakened its existence and eventually led to its extinction. This is similar to what happened in North and South America after their discovery by Europeans, European ships did not carry with them not only colonizers and invaders, but also carried deadly viruses and microbes, which European man developed immunity against them through exposure to them for thousands of years, while the Native Americans had no natural immunity against these viruses and with new colonists, over time, they melted and melted in the furnace of new societies. Created by European settlers and then disappeared forever .

Cannibalism theory
Many scientists have recently begun to believe that Neanderthals were cannibals, and that he loved to eat meat of his own species when hunting wild animals, fossilized bones found in some caves in Croatia, Italy, France and Spain had obvious traces of this. that they were chopped and shredded immediately after death, perhaps in order to eat them. The most compelling evidence of this brutal practice came from the El Sidron Cave in northern Spain, where scientists discovered a large number of bones, probably belonging to a small clan of Neanderthals that lived in this cave about 43,000 years ago and consisted of three women, three men, three teenagers, two children and one baby. Their bones bore scars, fractures, and marks, indicating that their owners were likely to be killed, dismembered, and eaten by another group of Neanderthals. The meat was removed from the bones with knives and stone axes, the skull was broken to extract the marrow, and the large bones were broken to extract the marrow .. All this indicates that the person who did this with these bodies most likely intended to devour them.
Scientists tend to believe that Neanderthals, like modern humans, repeatedly practiced cannibalism over their long history of tens of thousands of years, devouring their own species and devouring each other, and the frequency of this practice increased during

periods of disaster and famine, when food was scarce. , and cannibalism has been associated with superstitious rituals and beliefs such as those associated with cannibalism. The practice of cannibalism is undeniably brutal and terrible, but the worst part is that most readers may not be aware that it has serious public health side effects, and doctors may have been warned about it in their research and research into the causes of the strange disease. that struck members of the primitive Fur tribe in Papua New Guinea in the last century. This tribe used to devour the bodies of the dead, commemoration and mourning for the deceased included cutting and devouring by his family and friends in order to preserve the "vitality" contained in his body and save him from loss, where the immediate thought of what to devour the body of the deceased means sharing one's strength and energy. It was for this reason that men were interested in eating the heart of the deceased and the parts of his body that are most closely associated with strength and hardness, but the rest of the body was left to women and children, and they loved to eat fatty parts, especially the brain, they broke the skull and removed him to eat him raw. This strange practice has led many members of the tribe, especially women, to develop a rare disease called Kuru, a deadly disease similar to mad cow disease that is transmitted by infection and causes brain necrosis, turning it into a sponge. disease . It is also called the disease of laughter because the infected gradually lose control of their body, causing it to make involuntary movements that make them laugh. the patient usually dies in less than a year. there is still no effective cure for this disease, the incidence of which has declined after tribe members stopped devouring human flesh in recent decades.

Going back to Neanderthals, the team believes that cannibalism caused Neanderthals to develop serious diseases and epidemics similar to those of the Fur tribe in Papua New Guinea, which ultimately led to their extinction.
So, everything that we said about the Neanderthal remains in the space of possibilities of deception, scientists do not have information about this creature, which has become extinct only in order to offer its bones and Flint, which I use in pursuit of animals and enemies, and it is for this Because any lack of information is established, it remains the subject of a neanderthal topic thorny and controversial, especially among the supporters of the theory of evolution (1) supporters of the original detoxification (2) Life on this earth, which is controversial to the old stripper, we need evolution to him, because, honestly , we are not qualified not scientifically, but religiously, to go inside. But the essence of what we can say at the conclusion of this article is that the Neanderthals were neither an ape nor a monster ... and were not the ancestors of modern people .. Perhaps it was from creatures created by the creator before the appearance of people on this earth .. Nothing interferes with the hypothesis that the Earth was inhabited by other creatures and creatures even before the appearance of man

Crop circles (harvest)

There are many different opinions between supporters and opponents of this strange phenomenon, which began to expand and spread, and I think that most of you have heard of it, some of whom believe that it is man-made, while others believe that these are places where strange creatures land or maybe messages from smarter people? And we will learn more about this phenomenon.

Crop circles (crop circles) - this is a variety of geometric patterns and shapes are often produced at night and for a maximum of one night without being seen in the corn fields, wheat, oats and herbs. And it becomes clearer the further we go and look at it from above.

This phenomenon has become the most controversial topic among people and clients, as they have grown in complexity from year to year, even surpassed the analysis of scientists and attracted many people from different places, especially those who believe in the Paranormal and the supernatural.

Some people think they are made by aliens, some say they mean nothing, just crops of wheat and a lot of imagination, some see them as letters from flying saucers.

John is a farmer and farmer just like any other farmer who takes care of his vast field and the animals he raises on the farm. John wakes up early in the morning, collects eggs, milks the cow, eats his breakfast, and returns to work on the farm before dusk. John went to the farm one day after he had finished collecting the eggs and milking the cow, and if he was surprised by what he saw in the field, he found that a large area of corn sticks bent but did not break, but he didn't care because he knew that what had happened to his field of vandalism and falsification was due to some mistake. What happened to John was just the beginning of a series of big and strange circles that continue to this day with some farmers, including David, who grows grain on the Galtimore farm , but he is not happy at all because for the farmer that means a lot wheat that cannot be harvested, it costs a lot of money and gives nothing in return.

(David) suffers from a phenomenon that has been around for centuries this woodcut from 1678 called (Devil Harvest) shows the birth of wheat circles in the province of England this phenomenon returned again in the 70s and 80s, it was originally simple and small circles, but every day they became more and more complex and attracted more and more attention.

In 1991, all believers were disappointed that these circles were created by aliens when two people in the Southampton area of England, Doug Bower and Dave Charlie, appeared and

declared themselves responsible for the beginning of this phenomenon in 1978. They
made these circles for fun, to look like a landing site for flying saucers and aliens, and did
not expect people to react and believe so quickly. They became the most important events
in the news, began to appear in books and magazines, and people came to watch them.
from all angles. No one saw their creation, and speculation began to grow about the
identity of the creator of these circles, and so the enigma lived.

But there are commercial artists who consider this work as a profession for them, where
some farmers offer a piece of land to turn into a work of art at night, they create their art
collections, this method benefits farmers and artists with great benefit where tourists
come to see there are many companies advertising their products or services using this
method, many technical or television manufacturing companies, etc. such
as: Microsoft , Transportation Sweden , Nike , Pepsi , BBC,
(Andy Thomas) is the founder of the Southern Circle Study, which he has been studying
and writing about crop circles since 1991, says Thomas: "The first circle I saw was a good
distance away, something happened and I realized that there is a huge mystery going on
there and I felt it attracted me "many years ago Thomas watched hundreds of laps and did
a lot of tests, and then we learned that that night the shape we imagined appeared in the
area where we were doing test. There? We worked in complete secrecy and nobody knew,
as if someone was standing behind you and eavesdropping, accepting your idea and
applying it. "
Matthew Williams of the " paranormal believers" and "circle makers" says that the circles
that the count makes are a special phenomenon. before starting work, he calls on a higher
power to help him in his work. Matthew had things that he could not explain, and he did
not know how they happened to him. He did not find a convincing explanation. he once
created a circle that the believers said they imagined in their minds. He makes circles for
spiritual reasons, but they caused him problems, he was arrested once and fined after his
confession to make circles, and there is no doubt that planting circles have power over
human beings, rejoice and change their mood and heal them from some diseases,
according to some statements. Whatever happens, strange forces seem to interact with
each other to shape this mysterious event, and the first of these are new batteries,
cameras that break, and compasses that behave strangely.

Once he was in a circle near the farm of farmer David, when his mobile phone broke down,
he said: "I think it's weird because it just worked. What is this strange thing that happened
to Pavel and that energy that is not only science is still known? ..

Why aren't plant circles forming in front of witnesses? They appear when no one is
around.
"There is a very important reason why we suspect that these circles form at night: if they
are intended to attract attention and convey a message, then would they not be more
effective if they occur in front of us in broad daylight and in front of witnesses, why they
are always out of our sight, "says Ted.

Paul Piguet takes pleasure in reading the electromagnetic fields generated by plants and
this hobby led him to dedicate 12 years of his life to plant chains when he seemed
skeptical about it, except that over time he began to feel that there really is a strange

power emanating from these chains, and that it is not like other plants after study and study.

While Paul sees something extraordinary in these dimensions, Ted Clay considers them to be quite ordinary: "I am not surprised, it is natural that when wheat plants are straight, they look like little glowing sticks, and the electric charges are concentrated at the top, so they will differ from wheat plants leveled in the ground. "

Some fans of these circles say that no one can make such huge figures on a short English summer night. One of the most famous forms that people have spread is the one that appeared in 1996 opposite Stonehenge and appeared in no more than 45 minutes in broad daylight, which happened during a flight over Wiltshire . the pilot then landed at a local airport, refilled his plane's tank with fuel and took off again 25 minutes later, this time they saw huge formations of 151 circles, one of the most complex shapes of the time.

Now it's time to talk about the people who create the circuits and how the process of creating the circuit works.

One chain maker stands in the middle of the field while another walks around holding a rope or tape to take measurements that form the chain's perimeter, and when it comes time to decorate the interior, producers can do whatever their imagination dictates using the alignment roller. the ground as well as using your hands and feet.

Over the years, believing circle trackers, hoping to catch them at the scene of the crime, picked up night vision binoculars and checkpoints, but with all this, they failed to arrest a single suspect, which increased the hatred between the two teams.

Paranormal believers today believe that the crop circles have become very large for humans to make in a short summer night, which is 5 hours of the English summer night period, when a team of 3 tested it with the latest computer programs, they determined the angle and spiral formation and after completing the circle they follow them and the answer is simple to get to a point within the field. They should only follow the trail of the tractor and should never step on any standing plant, regardless of the circumstances. In less than 5 hours, the geniuses of nature completed their square and spiral shapes, and this shows that completing a circle can be done in one night, but this does not mean that all circles are made by humans.

Balls of light and their relationship to plant circles:

In July 1990, the photographer (Steve Alexander) decided to do his own research on farm circles, so he loaded the camcorder and started filming all of a sudden, he caught his attention with light floating, which could be interpreted as an optical illusion or even a glitch in the camera or even a white bird flying over the field. But there was an independent witness, a farmer who was driving his tractor, when he saw what he described as something glistening and sparkling flew overhead and then disappeared. what happened that day remains a mystery to this day.

The most influential and definitive piece of evidence for some paranormal believers was the one that emerged in 1996 and seemed to confirm the connection between balls of light and planting circles in a video showing a ball of light moving across a field to produce a magnificent shape, thus fulfilling everyone's dream. paranormal believer. This video was compelling evidence, but after a few days suspicions returned, and the controversy continues to this day. Regarding the balls of light captured on video, Ted says, "These things are called balls of light, but where does this light come from? Do you reproduce on your own? Or does it come from the sun , for example ?" One of the most important reasons he came up with is they are birds, scammers and things that the wind gets confused with, and as for the tape, many believed that Ted said he was surprised by the way of the photographer. Of course, you will be catching light balls, but the camera in this video is still not moving, and the sound of the operator is another matter. It's inappropriate What if it's just a gimmick and some believers still think the video is real. The harvest marks the end of the season of circles for both believers and skeptics, and no matter if you are a believer or a skeptic, one thing is for sure: it will return again next year, but do you still think it is man-made? Or are they from strange creatures who just want to powder our brains? It will remain a mystery that may one day be solved.

Weirdness of the Brain .. Turning Point

Change is the year of life and nature, everything changes over time, not limited to social, political, urban and cultural systems .. Even a person .. You and me, dear reader .. We are subject to changes at the level of personality, behavior, thinking and behavior, sometimes going from antithesis to antithesis. Usually this is associated with a certain event or experience, a turning point, after which everything becomes different from before, and there are plenty of examples of this .. He witnessed the death and funeral of his best friend, and he felt the fear of death and the frivolity of life, and he became pious, indistinguishable from a rosary and a prayer rug .. Flan , who was always joking, laughed cheerfully all the way .. He swallowed his tongue, and after the wedding, no one else heard him .. A neighbor's daughter failed her school exam, fell into depression and psychosis and in the end finally lost her mind .. Beautiful Valana , her young lover is dead ... And her heart .. Pledged to live in his memories until the last day of her life .. She never married and died an old maid.

These are all examples from real life .. I'm sure each of you has a lot of this story, but maybe it was some kind of model in itself for changing and wetting in the mysticism and magic of your channel, based on a certain life experience. Such changes and transformations are sometimes amazing and can be described as miraculous, especially when they happen so quickly and suddenly that a person turns into another person whose past does not matter. This is what we are going to talk about in this article .. Sudden and amazing changes in personality, which are a reason for reflection and reflection on how great the human mind.

He suddenly became a genius!

A few years ago I was watching a movie called The Phenomenon, a story about a simple mechanic named George (actor John Travlotta), who was standing in the middle of an empty town square one night and looking up at the sky when he saw a white dot moving between the stars. This simple mechanic woke up from a swoon and found himself a genius in all areas of science and life, his intellect became superhuman, he could absorb large amounts of information in a short period of time, he became able to formulate new ideas and innovations, and even had the ability to move things remotely !. It was later revealed that George's supernatural abilities were not caused by a miracle, as he thought, but by a cancerous tumor that somehow affected his brain and turned him into a superman.

The story of the film, of course, is fictional, but it raised more than one question in me about the possibility that a person could suddenly acquire such extraordinary intelligence and talents as George in the film? ... Imagine becoming a genius in physics, mathematics and computer science .. Overnight you become an outstanding artist in music, painting and sculpture .. Become a virtuous or compassionate poet .. Having the ability to speak more than one language .. What a wonderful dream, which has inevitably captured the imaginations of many of us .. But are they really verifiable? ..

Let's find the answer together in the following story ..

One day in 1994, surgeon Tony Scorea stood in front of a telephone booth to make a call, and by coincidence, lightning struck the booth from the sky at the very moment when Tony picked up the receiver and held it to his ear, so that the lightning bolt through the metal booth hit the phone and then into Tony's ear and brain, which immediately fell to the ground from the shock pit and his heart stopped beating. Tony later claimed that he experienced a spirit leaving his body in those few moments when his heart stopped, and said that he remembered standing next to the booth, looking with sadness and surprise at his body, lying on the ground, surrounded by blue a halo. Luckily for Tony, there was a nurse waiting for her to use the phone, and with first aid she managed to revive his heart and keep him alive until the ambulance arrived.

Tony recovered and left the hospital within a few days, suffered only minor burns and bruises, and suffered some memory problems for a while, but eventually bounced back and started his business as usual. However, the chapters of the strange story do not end there .. After a while, Tony felt an irresistible urge to play the piano .. He had no experience of playing the piano, he never sat at a piano in his life, and he had no musical interests. But suddenly and without warning, his head was filled with rhythms and melodies .. Even in his sleep, the piano played in his head non-stop!

Tony finally succumbed to his feelings and bought a piano, and what was his surprise when he sat down at the keyboard of this complex instrument and began to play without anyone's help, turning in just a few weeks into a dustless pianist! .. Music flowed from his brain to his fingers like a waterfall, and he himself could not interpret it. Naturally, this *sudden talent that fell to his lot attracted a lot of PRESS attention and a book was written about him. Today he is an outstanding pianist, invited to the oldest and largest opera houses and theaters in the world.*

Tony's story is really strange .. But it is not unique ..
In 2006, a similar story happened to a young man named Eric Imato , who hit his head on the edge of a pool while jumping into the water. ... Like Scoria , imato had never sat at a piano in his life, but a few days after the accident, he felt a strange attraction sitting at this instrument, and to his surprise, his fingers gracefully danced across the keyboard, as if he had been playing all his life. ... Musical notes flashed before his eyes, as if he were looking at a moving billboard.

"The blow to the head changed the mechanism and chemistry of his brain," says Dr. Andrew Revie, trying to explain the sudden talent of imato . Not only is this rare, but very unique. "

Truly amazing what a blow to the head can do! ..

About ten years ago, Jason Padgett was suddenly attacked by robbers when he walked out of a bar late at night. The attackers craved his leather jacket, and they did get it after they beat him in the head.

Jason lost his jacket .. But in return he got a supernatural talent in mathematics! ..
When the man opened his eyes again in the hospital, everything seemed different, Jason described it like this: "Everywhere I turned my face, I saw complex mathematical equations .. Like the Pythagorean equation .. Everything is curved .. Every circle .. Every tree .. It was part of the equation. "

Jason not only described what he sees, he turned his wonderful visions into complex mathematical schemes, which he called fractions. According to Jason, his fractals are: "Complex shapes, when you break them down into small pieces, all the pieces are the same or look like a whole."

According to some news reports, Jason is the only person in the world who can make a visual representation of Pi, a constant widely used in mathematics and physics. Defined as the ratio of the circumference of a circle to its diameter. Surprisingly, Jason was not interested in mathematics before the accident, was not a master of this subject at school and university, did not have any artistic inclinations. Today, his complex hand-drawn circuits have made him widely known and have been studied and analyzed by many engineers and mathematicians. Of course, this sudden genius caught the attention of doctors, neuroscientists, and brain scientists, and they subjected Jason to intense research and testing.

Dr. Bert Brugart , a neuroscientist and professor of philosophy, attempting to explain Jason's case, said: "It looks like Jason acquired his genius by pure chance after being severely beaten in the head by thieves, and with our tests we were able to diagnose intense activity in two specific areas of his brain that are responsible for arithmetic and mental imagery. The damage from the attack likely forced Jason's brain to compensate for his activity in certain areas and thus gave him access to areas of the brain that most people cannot access. "

Strange! .. But even stranger is that a person becomes mentally retarded and genius at the same time !! ..
Alonso Clemons was a normal child, like all children, but he suffered a serious head injury in an accident and this affected him greatly, his IQ went down and it became very difficult for him to speak, study and communicate with people, and he began to be considered disabled. But Alonso instead acquired a rare and amazing talent, or, as he says: "God gave me a gift." He could have made a statue of any animal, even if he had never seen it before, a simple and quick glance or photograph instantly turned into a three-dimensional mental image in Alonso's brain, which his fingers quickly translated into a beautiful statue in less than an hour. No one knows how Alonso does it, but over time his sculptures gained widespread fame and popularity, and some were sold for huge sums.

The wonders of the human mind do not end with Alonso's talent .. There is something stranger, more emotional and more human ..

The Englishman Tommy McHugh had nothing to do with art, no, he was the most distant people from art and beauty. An alcoholic known for his brutality and aggressiveness, while the streets of Liverpool complained about his orgies and fights, he spent most of his life behind bars on various charges such as theft, extortion, banditry, stalking and drug possession .. I continue to do this as long as he will not be fifty ..

Then the day came when he turned his being upside down.

One day in 2001, he was sitting in the bathroom, taking care of his needs, and someone seemed to knock on his door and pushed him out .. But he squeezed it so tightly that two arteries in his brain burst with it !. He had internal bleeding and passed out.

What a turning point! .. As befits a street criminal ..
After spending a week in a coma, McHugh opened his eyes and saw himself lying on the bed in a white room, and next to him was a woman who later told him that she was his wife .. He lost his memory! ... He also lost that dark, gray image with which his mind was used to seeing the world .. It completely disappeared and was replaced by a bright, magnificent image. As soon as McHugh got to his feet, he grabbed a paintbrush and began to transform this brilliant image of his imagination into beautiful, colorful paintings. Not only that, but also turned out to be a bandit who became addicted to swearing and swearing .. He turned into a sensitive poet talking about love, peace and hope, which, of course, did not like his friends, drug addicts, criminals and thieves, so they shook their heads ...

One of Tommy's doctors says: "Whatever the reason blocking or blocking his artistic prophecy, this thing broke and allowed the flow of emotions and images to open the door to the flow. We are still far from understanding the basics of brain functioning, but we hope that by studying rare cases like what happened to Tommy, we can get an idea of what's going on there. "

Tommy McHugh spends most of his time today painting, sculpting and writing poetry, even turning the walls of his house into vivid paintings, and he is as happy with his life as ever: "I love what I have become .. Then that I am .. I'm glad that now I am Tommy, and not the Tommy that everyone remembers .. What happened in my head is just wonderful. "

Let's hope that all the criminals in the world are squeezing their stomachs hard to rip their arteries and turn into artists and poets !! ..

But if Tommy McHugh has acquired a new talent and vision of the world, then there are those who have acquired a new language after the blackout ...

In 2010, a Croatian teenage girl woke up in a hospital after a 24-hour coma, to the surprise and surprise of everyone, the girl completely forgot her language. You can't speak Croatian anymore .. Not a single word! .. Instead, she began to speak German very fluently.
The girl's parents said that their daughter took elementary German lessons at school, but she did not speak this language at all, she remembered only some words and sentences, and only God knows how she began to speak German fluently after she woke up from a coma ...

"In the past, the girl would have been spoken of as a miracle, but today we prefer to believe that there is a logical explanation, although we have not yet come up with this logical explanation," says psychologist Dr. Migo Melas .

Indeed, history bears witness to many strange stories of people who, overnight, became native speakers of foreign languages about which they had no prior knowledge, a phenomenon that is classified as a supernatural phenomenon known as xenoglossia . There are very few documented cases of this phenomenon, such as the case of the Czech motorcycle racer Matthew Koos , who had an accident and lost consciousness for an hour, after which he woke up speaking fluent English !.

However, an explosion of talent and genius may not always require prolonged fainting or bleeding wounds. Orlando Cyril only accidentally hit a baseball in the left side of the head while playing with friends one day in 1979. Ten-year-old Arnaldo staggered and fell to the ground, but soon got up and played the match to the end. After the game, he suffered for several days from mild headaches in the left side of his head, but did not go to the doctor, because he believed that it was easy and not worth worrying about. When the headache finally went away, the dark-skinned boy found himself suddenly a genius at counting days .. For example, if someone asked Orlando about February 28, 1990, he would immediately reply that it was Wednesday and tell him that it was cloudy that day in Virginia, the city where Orlando lived.
What's surprising about Ronaldo is that his supernatural powers are only limited to the days after he was hit in the head in 1979, you can ask him what day is after that date and

he will answer you for sure, but his talent evaporates completely if asked what days before that date.

Arnaldo told scientists who have studied his condition that he cannot explain his talent, it is like watching TV, as soon as someone asks him about a specific date, until he finds the name of the day corresponding to that date embodied before his eyes.

He suddenly became a criminal!
In 37 AD, Caligula took over the throne of the Roman Empire, and joy spread throughout the Empire. The people were very pleased with their new emperor after they tasted bitterness and humiliation at the hands of the former emperor Tiberius, who burdened the people with his taxes and insults. The young emperor soon proved his worth with the love of the masses, with tax cuts, reforms and much discontent. Just seven months after his accession to the throne, Caligula fell ill with a mysterious disease that doctors could not cure, and was so ferocious that everyone despaired of his recovery and was convinced of his death. But contrary to all expectations, Caligula surprised everyone and recovered from his illness. However, this is no longer the person he was before the illness, his character has completely changed, he became a sadist, greedy and punk-annoying, he kills people and appropriates their money, he will rape their women for no reason, and in Wertheim's debauchery he is that he did not leave the woman in the tile only and raped her, did not betray his three sisters. Moreover, he opened brothels in his palaces and forced the Roman nobles and overseers to pimp their wives and women in them! .. The proceeds from these brothels, of course, went into Caligula's pocket. If only this ended, Caligula's madness would be incredible. The same statues and temples all over the country made people worship him. His army took a tractor to the shores of the Manche Sea between France and England to collect their shell soldiers on the coast, and then brought them back to Rome! ... He again imposed heavy and burdensome taxes on the people, until the people had mercy in the days of Tiberius, according to the proverb: "Judge others. You know my mercy."

But luckily for Rome, his odious reign did not last long, as he was killed by his guards in 41 AD. His story is just a historical example of the phenomenon of personality change for the worse after an illness or accident. People don't always become geniuses after accidents, as in the stories we've already mentioned, but they can also turn into eccentrics .. Bandits and criminals who love blood and torture others.
In 2007, a Russian court sentenced to life imprisonment Alexander Pichushkin , aka the chess thug, who lured his victims into the nooks and crannies of Moscow's Petsovsky Park, where he treated them to vodka while they weren't getting drunk, and then beat them on the back of the head with a deadly ax. Sometimes he liked to dip a bottle of vodka into their broken skulls and then leave them in the open without burial. Pichushkin was an experienced chess player who from an early age came to the park and played with the old people, pensioners, unemployed and homeless people who were there, and in fact they made up the bulk of his victims. He hoped to kill 64 people by the number of chess squares, so he was called the chessboard ripper , and he would have been very close to achieving his goal if he had not been arrested in 2006.

Pichushkin confessed to the police about the murder of 60 people. He told them that his first crime was committed in 1992, the last a few days before his arrest, and his colleague was the last victim. When asked about the motives of his crimes, he replied: "For me, life without murder is like life without food for you. I felt like the father of all these people, I am the one who opened the door to another world for them."

Pichushkin 's mother said that the real reason for her son's crimes was that as a child he had an accident when the presenter hit his head on a swing, and his soft and calm nature changed, he became aggressive and hot-tempered. Experts believe that the frontal cortex of the brain, that is, the area where Pichushkin was injured , is responsible for suppressing anger and hostile tendencies.

The Chess Thug was not the only one whose life turned for the worse after being hit on the head. American serial killer Earl Nelson was ten years old when his bicycle collided with a tram, he was hit in the head and fell into a coma for six days, and when he finally regained consciousness, everyone noticed that he had changed completely. His actions became strange, he suffered from headaches and memory loss, he even went to a psychiatric hospital twice. This obsession gradually turned into an anomaly, and he became addicted to sex with dead bodies (necrophilia), which led him to take the lives of many women along the west coast of the United States and Canada.

He pretended well, his victims were mostly boarding house owners, he walked into their homes claiming to rent a room and always held a copy of the Bible in his hand to make a good and calm impression of himself, and this trick always worked as soon as the victim will calm him down and give him his back so that he will jump to her like a predatory beast, putting his hands on her neck and leaving her only a dead body, he will make sure that she is dead before he starts having sex with her, and as soon as he empties it so he can pull it out and hide it under the bed. Hiding corpses under a bed was a hallmark of the crimes of Earl Nelson, who killed more than 22 women in this way before being arrested and executed by hanging in 1928.

Fritz Harman is another example of what a blow to the head can do. Harman , known as the Hanoverian Ripper , was a vampire in every sense of the word who hunted homeless boys from the streets and brought them to his apartment to rape, and at some point during sex, he bit and cut his victim's throat with his teeth, and then he began to drink blood directly from the bleeding wound. After the death of the victim, he cuts the body and removes the meat from the bone in order to sell it on the market as pork !.

Harman was not a born murderer, he was a soldier in his youth, for whom ambition and perseverance were known. But one day during a training exercise, he fell from a height, hit his head and lost consciousness, and his colleagues agreed that his personality and behavior changed dramatically as soon as he woke up from his coma, and since then his life took a different turn, which eventually led him to the guillotine, which was beheaded in 1925 for the murder of more than seventy people.

In fact, there are many people whose lives have changed towards violence and crime after suffering a head injury, and if it were not for the fear of prolonging and clarifying, We would just take a few samples. :

- French serial killer Joseph Vachier was a specialist in the rape and killing of sheep and cow shepherds in the wild. As a young man he suffered serious head injuries from a suicide attempt.

- American serial killer Albert Fischer .. Pedophile and cannibal .. As a child, he fell from a tree.

- English serial killer John Christie .. Rapist and murderer of women .. On one of London streets he was hit by a car and lost consciousness for several hours.

- Killer Richard Speck ... In 1966, he broke into the internal ward and killed eight nursing students .. As a child, he suffered many injuries, and when he died in prison in 1991, doctors opened his brain and discovered something amazing: a region, responsible for memory intersected with the area responsible for anger and strong emotions, the border between these two areas completely disappeared.

- Gary Hayden , you ... Rapist and murderer of women ... He fell from a tree as a child and his wounds were so severe that they affected the shape and consistency of his head.

- Randy Kraft .. The serial killer who raped and killed dozens of boys ... He fell on the stairs in his childhood and passed out for several hours.
- Bobby Joe Long .. Rapist and killer of women .. He had many head injuries, once he fell off a swing, once he was hit by a car. He had an amazing love of sex, he raped over fifty women and killed over ten, he once said, "when I die, they will open my head and they will find exactly what I always told them ... Part of my brain is black dry and dead. "

- Fred West .. The rapist killed more than three dozen women with the help of his wife Rosemary, three of his victims were his daughters .. He fell off a motorcycle at the age of seventeen, hit his head and passed out for a week, and his family said he was violent changed after waking up from a coma.

- Richard Ramirez ... Devil worshiper , he killed more than 14 people and raped many women who were horribly represented by the bodies of his victims .. As a child he suffered twice head injuries, once a wardrobe fell on him, and once he was hit by a swing.

Many other criminals have suffered head injuries at some point in their lives, and these injuries have directly impacted their lives, altering the machinery of their brains and changing their perception of the people and society around them.

Does this mean that the brain of criminals is different from the brain of other people? ... For many years we have been accustomed to saying that it is society that makes the criminal, and that the environment, living conditions, education and family play the biggest role in creating criminal tendencies. Poverty, for example, is one of the main causes of crime, and dysfunctional family life is a powerful factor in crime among children and adolescents.

But it should be noted that not everyone who grew up in broken families becomes murderers and thugs, not all poor people can turn into thieves, and not all disadvantaged women end up in prostitution .. There is always a certain percentage of people who are ready to do something, regardless of family background , cultural and physical level. It's about genes and brain structure. Recent scientific research has shown that the brains of criminals are actually different from the brains of other people. Scientists took a sample of criminals in prison and performed MRI scans of their brains, and found that psychopaths, who make up about a third of the total number of criminals in the sample, had less active and visible gray areas in their brains than in the brains of ordinary people, and scientists it is said that these gray areas or spots in the brain are responsible for dysfunction in these areas, which has a great impact on the conscience of the offender, as he never regrets his shameful actions towards others, does not feel guilty for harming those closest to him and is not responsible for its mistakes. Feelings and sensations of others for him do not mean anything, because he does not even feel them, only he is important .. This is absolute egoism. The danger in the owners of these sick characters is that they do not look like sick people, they know how to control their emotions, they know how to pretend to be in love, sympathy, tenderness, they know how to lie and deceive without the slightest remorse, and some of them can rise in society to titles of rulers and leaders.

The same can be said for Antisocial Personality Disorder, which also makes up a small percentage of the criminal population. Scientists believe that genes, hormones and nerves play an important role in the creation and shaping of this dangerous personality.

So our mind controls everything? ..

Yes, that's true, and our minds are different, like fingerprints and eyes are different .. But the biggest role has to do with the nature of our genes and the composition of our brain, which is why we find that intelligence varies from person to person, there are geniuses, such like Einstein, who can solve dozens of difficult math problems in minutes, and there are people like me who have a headache when they ask them for the sum of four numbers !.

Black hole .. Hidden monster of the Milky Way !!

The black hole is known to scientists as a dead star with a very, very, very massive gravitational force !! So nothing can escape from him, and his destiny is to be swallowed up by this hole, like black holes move without stopping.
But what is the secret of this monster's attractiveness?
The secret of a black hole's gravitational force lies in how very, very dense its mass is, which gives it such a huge gravitational force that no object can escape from it, not even the light that can reach from the sun to our planet Earth in just 8 minutes .. Even this light cannot escape the black galactic monster.

In fact, a black hole is not a literal hole, it is also not empty, it is full of material concentrated in a small space, which gives it a tremendous gravitational force, and black holes are called that because they look like that in space.

I mentioned in the title that the black hole is a hidden beast of the Milky Way, and it really is because we cannot see it, because it does not emit light, but when it attracts another star, it becomes clear at that moment, and this is due to the plasma beams and their rotation around it.
Is there a black hole at the center of our galaxy?
Yes, there is a black hole in our galaxy, and it was spotted by NASA, and its mass is about three million times the mass of the Sun, and it is about 24 thousand light-years from the Sun, so it is too far from the earth to be a danger to it.

There is a question I have often asked about black holes, which is, since black holes are dead stars, could our sun ever turn into a black hole?

And the reason is that the sun is too small to become a black hole, for a star to turn into a black hole, its mass must be so great that it collapses and becomes a black hole, and the star does not turn into a black hole until then. until the material inside the sun is confined and subjected to superhuman pressure, and all this only happens if it has already turned into a black hole or one that is close to becoming a black hole. The smallest stars that turned into black holes were 20 times the size of our Sun or more, and if you think that our sun is huge, here is a video and you will get to know the largest types of stars discovered so far, and compare them with the Earth and the Sun.
Okay, but what happens if two holes meet in front of each other?
Some will probably say that they will swallow each other and they can explode, which is a logical answer, I said at the beginning of the topic that a black hole swallows everything that comes in its way, but the answer is incorrect, and if this happens, two black the holes will meet, they will merge and become a very large black hole that sweeps away everything in its path.

Types of black holes
So far, 4 species have been discovered, and, in my personal opinion, there are many more. These four types are:

- Micro-black holes: These are considered the smallest species and scientists call them quantum mechanical black holes and are believed to be the oldest type of holes.

- Stellar black hole: It is similar to the one we talked about earlier, to the one that results from the explosion of an aging star.

- Medium mass black hole: This type is much larger than stellar black holes, but much smaller than the star that we will now talk about.

- Supermassive Black Hole: This massive monster has billions of times the mass of the sun and is formed when two holes merge with each other, as we said earlier.

What is the evidence for the existence of a black hole ?!
After the discovery of the black hole, Western scientists celebrated this brilliant and mesmerizing discovery and were about to take off with joy, but they did not know that the creator of the heavens and the Earth, the Lord of the worlds, created this cosmic miracle, perhaps a dear reader will believe in his existence if he reads one of verses of the mighty book of God about heaven and earth.

The verse testifying to the veracity of the existence of black holes is the saying of the Almighty {do not swear by the khans-neighbors of sweeping} granulation 15-16.

Do you know what neighborhood sweeping is ?!

These are black holes .. And their interpretation:

1 - and so on: everything that disappears and is never seen was called devilish diphtheria, because the children of Adam did not see it. This is what scientists use the word invisible, that is, invisible.

2. such as: that is, it works and moves at high speed. This is what scientists express by the word "move", that is.

3-Sweeping: That is, sweeping and absorbing whatever you come across in your path. This is what scientists express with the word vacuum cleaner, that is, a broom.

Oddities and wonders of disease

Health is a crown on the heads of healthy people, this is a phrase most likely said by a sick person, healthy people rarely pay attention to it .. And why do they care about what they already have ?! .. One of these healthy people, an unemployed young man, once commented on this phrase: "What is the use of this imaginary crown? .. I need real gold. Give me a house and a car and take the crown from me." One rich patient looked at him with a broken heart and said: "I am ready to do whatever I have for this crown. And you replace it with a machine!"

He ran back and forth into the bathroom, suffering from excruciating pain, until he fell into despondency and tiredness, crying and begging: "Servants, army, coming ... Help me, I want to pee! I will give my whole kingdom in exchange for urination." This quote later reached the ears of one of the sages, who sarcastically remarked: "The ugliness of God is the king worth his urine!"

Yes, dear reader .. When illness, even the greatest and best things, lose their brilliance and splendor, a sick person can do everything that he has for his health, but correct recovery forgets that there is grace in him .. Money. Clothes ... Work with the nose! .- Anything that comes to mind. While the patient asks only one command ... Every night, after his eyes fall asleep and faint from the length of his legs and rest, he stretches out his palms to the sky, washes his cheeks with tears, trembles his lips with hope, prays to his creator, crying and begging: "Lord , I touched evil, and you are the Most Merciful of the Most Merciful ... ".

And what a piercing and sad sight ... The sight of these madons enveloped them in the sky, they prayed.

But weren't these same people once healthy? .. There was no one unjust, mighty and arrogant in them .. Look how they got .. Look at this cruel father how he became a meek Lamb who flatters his children after he got sick. And this rich, haughty man, whose feet almost never set foot on the earth because of his arrogance and imagination .. Look at him, how he prays for the prayers of the poor! .. And this Haifa Gada who is addicted to breaking hearts .. She became lonely after her beauty was gone and a crowd of fans and lovers were with him ..

God does not gloat in sickness .. But the various states of people when they are healthy and sick are what requires meditation, and this is a sermon and lesson for each consideration. The other day I went to the clinic to see a relative who had a malignant head disease, and I had not seen him for a long time, and he was entrusted to me by a strong young man. I knew he had several surgeries and was treated with chemicals and I saw him and his body disintegrated, his tongue felt heavy, his eyes turned, his hair fell, and his mind brightened. I thought I would write an article on the strangeness of disease, in which I would warn the healthy and not pay attention to the fact that there is grace in them, and entertain the sick by exposing the misfortunes of others.

Alice in Wonderland

You may marvel at this name, dear reader! .. What does Alice and her interesting adventures have to do with the world of pathology? .. But believe it or not .. There is actually a disease called Alice in Wonderland syndrome, and the reason for this strange name is related to the sensory hallucinations that accompany this disease, so although the patient's visual system is completely intact , his vision of things uneven .. The ferry cat is the size of a car .. And the controls are the size of a cat! .. And his friend's head becomes the size of a lemon and a palm the size of a skyscraper. The sufferer may have trouble determining distances and the relative will soon see far and wide. Not being able to distinguish colors, he creates his own perception of reality, the file on which he focuses, although he walked slowly, or he is nailed to the place and unable to move. In some cases, visual impairment may be due to impairment of touch and hearing, and nearby sounds seem to come from a distant location.

The infection occurs most commonly in adolescents under the age of twenty, rarely used under thirty. It occurs in patients with migraine, cancer and viral brain lesions, and also occurs with excessive use of drugs, alcohol and sedative pills. The attack may recur several times a day and take some time to calm down. In fact, it is harmless and requires only rest, but its effects strongly affect the patient's psyche due to the horror and anxiety that it causes.

One teenager with this syndrome describes how he feels: "Suddenly things seem small and far or big and close. I feel like I am shorter, that my size has decreased, and I see other people as if they have become dwarfs. Sometimes I see a curtain on an open window, or a TV turned upside down, or my arms and legs dangle. People's voices can get too close and too loud or too weak and too far away. "

Allergy to water

Dear reader, do you believe that there are people who are allergic to water ?! .. Their contact with any kind of water has serious consequences, even their tears are harmless! This unique condition, called aquagenic urticaria, a type of skin urticaria that is caused by the body's sensitivity to a certain substance, appears as red spots accompanied by itching and pain and can last for hours or days, and is called chronic if it lasts more than six weeks. Even the tears and sweat of the sufferer lead to symptoms, and for this, sufferers try to always stay in cool places and not sweat as much as possible. The pain and burning sensation caused by tears on the eyelids leads the sufferer to a continuous cycle of groaning. Sometimes even drinking water can be painful and heavy, causing throat swelling and shortness of breath for ten minutes or two hours.

The worst aspect of this disease is personal hygiene.Therefore, bathing is a terrible nightmare for the injured, they cannot wash standing in the shower or lying in the bath like other people, because of this, they have serious consequences, and for this they resort to cleaning their bodies with a sponge, moisten them with a small amount of water and slowly and carefully pass them through different parts of the body, mixing with others, fearing to be accused of uncleanness and not taking care of their personal hygiene, most often People will not believe that such a disease exists like a water allergy.

Finally, urticaria is very rare, only forty people are infected worldwide, and unfortunately there is no viable cure for the disease. The most effective treatment is prevention, that is, staying away from water and not sweating or crying.

Premature aging
A rare and strange condition known scientifically as Progeria Syndrome , or premature aging. There are only fifty or more cases of the disease worldwide, and the disease is so rare that it affects only one in eight million newborns.

People with this disease are born naturally, the symptoms of the disease appear only after a few months, and the first symptoms are slow growth, skin aging. Then the symptoms develop steadily, hair loss, joint stiffness, facial features become characteristic of a small chin and a pinched nose. The older the poor child is, the more he looks like old people and the more severe the illness on him, wrinkles his skin, weakens his vision, his head becomes bald and gets illnesses and injuries that usually only affect the elderly, such as kidney failure, atherosclerosis. A child often dies of cardiac arrest at an early age, as children with progeria are only 13 years old and can last up to 20 at best.

The disease affects the shape and characteristics of the child, but does not affect his mental capacity. This is not a genetic disease, that is, the mutation (genetic defect) caused by it is unlikely to recur in the patient's relatives. Unfortunately, there is still no effective cure for the disease, despite efforts to find a cure, or at least one that could curb the rate of the disease.

Elephantiasis
This is a terrible disease, but it is not uncommon, there are about 120 million infected worldwide, especially in hot areas. But the disease, fortunately, is not fatal , rarely leads to death.

Infection occurs due to a microscopic parasite carried by mosquitoes from the body to the corresponding organ, these worms used in the lymphatic system not only live there, they usually concentrate in the lymphatic vessels present in the lower extremities of the body, especially in the leg and foot. Reproducing and growing to a large size, these worms block the lymphatic vessels and accumulate lymphatic fluid in them, causing the surrounding tissues to swell and deform. Since the disease usually affects the foot and causes it to swell up to gigantic proportions, they called it Elephant Disease.

The disease, as described above, concentrates in the lower extremities, that is, in the leg and foot, and is less common in other parts of the body, but can affect the male reproductive system, so the testicle becomes larger than a basketball. It can infect a woman's breasts by turning into a giant misshapen sac that hangs down to the ground. In rare cases, injury to the hand or face may occur.

The reason for the slow treatment is the doctors' fear of killing the worms right away, so that their fragments do not accumulate in the lymphatic vessels beyond the body's ability to get rid of them, causing poisoning, leading to the death of the patient. Treatment can sometimes require surgery, neglecting the patient to treat his injury early because the tumors are painless, and because of poverty and ignorance, especially in poor and remote areas. This negligence causes the affected parts to swell up to giant sizes
.. And then only surgery will help her .

Tree man

In 2007, millions of people around the world were surprised to see images
of Didi Kuswara . ... I slept on his body, especially on his arms and legs, the growths of his skin resembled the bark of trees. Then a rumor spread that he was a sinful man, whose indignation God had led to a tree! ..

A man is just a skin disease warty dysplasia of the epidermis
(Epidermodysplasia verruciform). This thick, bark-like layer on his skin is actually just warts, like the ones that grow on most people's bodies, but Didi's problem is that his immune system has a rare type of defect that makes him unable to withstand these annoying warts. and they grow uncontrollably to record sizes. She really doesn't hurt like all warts, but her impact on Didi's life has been devastating since her first appearance due to a wound that hit him in the leg in the woods in his prime, because of which he lost his job and his wife left him alone. with two teenage daughters , and since no one hired him, he was forced to work. And as a harmful and useful gentleman, as they
say, Didi's warts irritated him, became a means of earning his strength, since his work in the circus led to his reputation, vtahftat he had newspapers and satellite TV to photograph him and publish his story, and then they said that an American doctor had found a successful treatment for his condition, and that But it was just a lie, like all the lies that revolved around this man.

Diddy did undergo surgery in America, during which doctors removed 95 percent of his warts, but they soon grew back. He had a second operation in America and another in Indonesia, but each time the warts grew back after a short time. The Indonesian government then banned Dede from traveling abroad on the grounds that the Americans wanted to turn him into a test rat. All of Didi's operations were carried out free of charge and at the expense of charity.

Didi himself , who still works in the circus, says that he is ready to go for treatment anywhere in the world. Didi's case is extremely rare, with a probability of only one in a million, and only three cases are known worldwide.

Bubble man

If Didi was known under the name of the tree man, then there is another Indonesian man named Chandra Visnow , better known among the people as the Bubble Man. This name is associated with hundreds of small blistering tumors that cover his entire body, the result of a disease called type i neurofibromatosis , a genetic disorder that affects one in 2,500

*people worldwide, but Chandra's condition is considered an extreme form of the disease,
in most cases mild to moderate ... This defect leads to the loss of a protein
called neurofibromin , a protein that helps regulate cell growth, and its loss leads to
uncontrolled cell growth, which leads to the appearance of benign tumors in various parts
of the body.*

*Chandra says that the symptoms of the disease first appeared when he was 19 years old
and that tumors covered his entire body by the age of 32. Unable to diagnose and find out
the cause of their illness for many years, local doctors desperately turned to traditional
healers and sorcerers, but to no avail. He was diagnosed only after he turned 50, an
American doctor who volunteered to help him after seeing his photographs in the
newspaper. Unfortunately, Chandra's disease is not cured even in developed countries, but
there are medicines that can alleviate the disease and improve the patient's life.
Chandra's tumors are not painful, but they can cause an unpleasant itch so intense
that Chandra only feels comfortable after turning off the butt in areas that bother
him. The second bad aspect of Chandra's illness is the expression of disgust he sees in the
eyes of most people when walking down the street, but he has found many ways to deal
with his illness, and this is what he says:*

*- I always try to occupy myself with something in order to forget what I am in. I try not to
look in the mirror, so as not to remember my tragedy .. And when people look at me on
the street, I tell myself that they are looking at me because I am handsome .. Always try to
look on the bright side. "*

*Today, Chandra is most worried about the fate of his son and daughter, they are already
in their 20s, and recently they have developed symptoms of the disease, but the doctors
reassured Chandra , although they cannot cure the disease, but they are able to prevent it
from reaching the same severe stage as at Chandra's .
It is worth noting that neurofibromatosis may not cover all parts of the body, sometimes
single tumors appear that are concentrated in a certain area of the body, but can grow to
record sizes, as happened with the Chinese Huang Gang, nicknamed the Elephant Man,
Who surprised the world when he the story was first covered in 2007.*

*Huang was born in 1977 to a poor Chinese family, and his parents noticed a swelling on his
face when he was four years old, took him to the hospital and diagnosed him, but doctors
did not advise removing the tumor with surgery due to the severity of the operation at the
time. Thus, the tumor continued to grow for many years until it became gigantic and
completely deformed Juan's head. But as medicine developed, Chinese doctors risked
performing several complex surgeries for Huang in the period 2007-2008, during which
tumors with a total weight of more than twenty kilograms were removed.*

*The real daddy of Smurfs !
When children see him on the street, they run to him, shouting joyfully: "Papa Smurf !" ..
He greets them with a smile .. But you shouldn't be on a par with the big boys who dare to
call him by that name, and feel comfortable with the tracking dogs that chase him
wherever they are ..*

This is Paul Carrison , the Blue Man that most people didn't believe existed until he appeared on American television (today) in 2008. Even after his appearance on television, many thought it was a hoax, people were not used to seeing people with blue skin, except perhaps in science fiction films such as " Avatar " or in the cartoon "The Smurfs ".

It is diagnosed and known for a long time, known as argyria , and it is acquired, that is, carison was not born with blue skin, but acquired this color as a result of his use of silver-containing preparations.

This whole story began in the early nineties of the last century, when a chronic rash appeared on Kerrison's face , and he tried to treat it with a colloidal solution of silver, which he himself made at home by electrolysis, and continued to use this solution for more than ten years, as a result of which silver was deposited in tissues. his body, oxidizing it to black through oxidation.
Before the discovery of penicillin in the first half of the last century, silver was used as an antibiotic to treat infections and disinfect wounds. Certain silver-based products and medicines are still widely available in many countries around the world, but have been banned in the United States since 1999 due to their side effects, especially skin discoloration that can affect the skin when products containing silver are used for a long time.

Paul Carrison died in 2013, and his death had nothing to do with the condition of his skin, he died of cardiac arrest.

Finally, while searching for sources on this topic, I noticed that many Arab and foreign sites and forums recommend the use of a colloidal silver solution and attribute miraculous properties close to miracles to it .. It cures all diseases and cures all types of cancer !! .. Of course, I am not a doctor and I don't understand anything in medicine .. But with the help of reason and logic .. If all these miraculous properties were present in a solution of colloidal silver, why would doctors in the West replace it with penicillin and other modern antibiotics .. Really its advantages are absent from them and from those who have used it in their hospitals for decades ?! ...

Drowsiness disease
Many epidemics have struck humanity over its long history, some of which have been engraved in the memory of people for centuries, such as the Black Plague and smallpox, which killed millions of people in the Middle Ages, others that went unnoticed, are barely remembered, such as the disease of sleeping encephalitis (Encephalitis lethargica), also known as the disease of sleepiness, grew into a global epidemic between 1916-1926, leaving behind many others. I think that the nature of the disease and its appearance in an era full of disasters such as the First World War and the Spanish flu epidemic of 1918, everyone hastened to forget about the disease, rarely find anyone to mention it, and most people confuse it with sleeping sickness caused by the Tsetse fly in Africa. Moreover, after the 1920s, the disease completely disappeared, no longer struck like an epidemic, and new infections were limited to isolated cases, numbering no more than a dozen worldwide.
So this is a deadly epidemic that has been forgotten, even doctors do not want to remember it, it is a mysterious disease that makes them sick and has a headache, because

they cannot know its causes or how to cure it. He is strange in his roles .. It starts with pain in the oropharynx, followed by fever, headache, and well, fatigue .. Then this condition develops into sleep disturbance and muscle spasm .. Finally, the patient enters a strange state, reminiscent of a coma, but non-coma! ... Rather, he turns into a statue .. He doesn't move voluntarily .. And don't talk .. But he didn't die .. It's a prisoner inside his body .. Alive Dead .. It can last for months and years and rarely heal.

The disease is inherently associated with brain infections. But the most amazing thing about this is that doctors did not find the cause of this inflammation, they examined hundreds of samples from the brains of patients, but did not find any viruses .. Why, then, does the brain become inflamed ?! .. This is a question they have not found an answer to.

One recent study I read about briefly says that the cause of the disease is a rare type of bacteria known as streptococcus, which doctors have found in most patient samples, and it triggers a violent immune response in some people to this extent. that their immune system attacks cells and tissues.
The disease remains incurable to this day. But in 1969, there was a strange and poignant incident at Beth Abraham Hospital in New York, and that same year, a young doctor named Oliver Sachs decided to test a new medicine for Parkinson's disease on some carotid encephalitis patients who were among the few survivors. the epidemic that hit them in the 1920s after being hospitalized for decades. The new drug was called Levodopa , and it did something like a miracle, the patients regained consciousness and returned to normal after taking the drug .. They were like living statues that suddenly began to move .. Unfortunately, this did not last long, since all patients soon relapsed and returned to their first state, and the doses of the drug no longer worked on them. Nonetheless, their story was incredibly moving, so it should come as no surprise that it was taken up by Hollywood in a film called Awakening, in which Robert De Niro presented one of his most beautiful and wonderful film roles.

Sasquatch (Bigfoot)

Bigfoot is one of the most mysterious mysteries of all time. Or a gorilla that has some human traits? .. Maybe an extinct person? ?...

In 1967 in Northern California, American Roger Patterson was riding a horse with a friend, enjoying the picturesque views .. Suddenly, they were cut off by a huge creature! ... Hairless, human and gorilla look like a female body! ..
The sight of this strange creature frightened both men and frightened Patterson's horse, throwing it to the ground, while the strange creature fled, hiding in the trees. But Roger Patterson pulled himself together and quickly got up to turn on the camera, which, fortunately, he had at that moment .. The creature caught the short film, being far from this place. Roger's friend Bob Gemmellen noticed that the footprints of the creature were deeply cut into the ground, which indicated the weight of the mysterious creature, so he made a plaster template for the tracks .. This blow was several times larger than an ordinary human leg .. Hence the name "Snowy person".

It is noteworthy that these were not the first sightings of the creature .. And certainly not the last .. Every year there are many reports that we saw this freak .. Especially in Northern California, where more observations were obtained than in any other area! .. Many of these views may lead us to wonder: what makes all these people lie and fabricate such stories? .- And what is the benefit to them?

And in order to verify the fact of the film that Patterson directed, it was shown to technicians and specialists to expose his lies from his honesty! As a result, some of them saw the content of the film as Real, while others saw it all as a lie and said it was nothing more than a gorilla costume. The bottom line is that everyone agreed that the film was real, but they disagreed with the reality of the object that appears in the film .. Is it really a mysterious unknown object or a person disguised as a gorilla.
With the advent and distribution of Patterson's film, there were hundreds of Bigfoot supporters and fans ... claims and stories abound that he watched him ... Some of them are probably real! ... Most of them are just a bear or gorilla that someone saw and was very happy.

But none of these stories have attracted as much attention as the Guinness Factor, a story that cannot be ignored, as one researcher said .. And this Guinness woman is in her thirties .. Simple dirty looks, lives a quiet life in the countryside .. Life may seem ordinary and insipid .. But it is not! ...

Guinness claimed that she grew up in a family full of big-horned people, and that she knew their little ones from their big ones, and the details of their life .. How they eat, sleep, etc. They first started visiting her grandfather 60 years ago. and it was there that a long and close relationship developed .. And what kind of relationship.

Janice talks about one of their visits: one day Bigfoot knocked on my door, and when I opened it, I saw that he was asking me for some garlic ... Fortunately, the other day I had some garlic and I gave it to him .. His voice was very deep! A little scary and restless .. When he says "hello", for example, he says it like this: Go ..BA.

Janice said that and smiled tenderly, as if she remembered her cute pet or something.

But how do they stay alive? ...
This was one of the questions that Guinness asked and she was very happy to answer it, she claimed to have seen them several times on a fishing trip, yes, they hunt at different times and also alternately. They run faster on 4 limbs than on two, so when they start hunting, they run on 4 limbs and chase the deer until they get to him .. And when this happens, they grab his hind legs, and he falls on the ground, they break his neck, and then, using their nails, which are very sharp, they rip open his stomach and extract his intestines, heart, liver and so on ...

Janice fell silent, I thought a little and looked at the researcher with irritation: (when I was young, it annoyed me, but now it worries me, I don't want to watch them kill someone!).

Janice understands perfectly well that from the point of view of the world she is just another naive person living in a fantasy world, but for her big-mouthed people are a reality, and they are her family, and she loves them so much, they are very smart, they have culture and tongue and ... a little shy in front of the camera! ..

What if it's a prank?

With this possibility in mind, all possibilities cannot be ignored in the end and with reference to the evidence .. The only proof they have is Roger Patterson's film. "

Only 3 people are witnesses in the film .. "Roger Patterson" is dead .. Do not forget that their search for the truth in this case went on almost 50 years after the film was filmed, so Roger must have been dead and numbered among the face of the blessed long ago .. The second witness is Bob Jimlin . the other never talks about it .. And since they see it as a hoax, there must be someone under a Bigfoot costume who was under that shock of hair? "Girls" is the name of an old friend of Roger's, and Bob often traveled with them, linking him to a close relationship .. This hieraln so strongly claims that he was behind the costume! He said it for a reason .. And, of course, there are many who believe his words and support what he says, he also said that many people know about his disguise, but they were silent.
" Geralnos " said Roger offered him a deal that should never be ignored, in exchange for wearing a suit within 10 minutes he will receive from the hands of a thousand dollars, he was twenty-six years, he had a lot of free time and a little money, so he took the best advantage of the show, agreement and practice followed, they all shook hands and left.

See the "areas" of what happened :(we finally arrived in the region where the film will be filming, they help to wear the uniform as it was a little heavy and hot and warned me "Roger" from some hunters pregnant shotguns Lost me, and they leave holes that are a lot in my body !! They asked me to stand in a certain place and I followed their order, and then I went and went and I don't know how much I was walking, I looked at them during filming, and then I continued on My path, and I am sure that someone is going to shoot at me, and he is happy that he is hunting a stranger, and ... But it was all over, and I felt that I was fine.

(If they had paid me the thousand dollars as promised, I would have kept my mouth shut.)

So the Geraln ended his speech , and a faint smile appeared on his face.

But .. No one has ever found the suit .. And there is no evidence .. And why will all the inhabitants of his village be silent? Have none of them dared to expose Roger Patterson all these years?

Explanations:

The possibility of hundreds of people claiming Bigfoot existence remains elusive. That is why the question arises.

What if Bigfoot really exists? How to explain this and where did it come from?

One theory says that Bigfoot is a descendant of one of the giant monkeys that went extinct a long time ago! And the fact is that he is just a humanoid monkey .. Or is it an ordinary monkey, which over the years has turned into a product of a small monkey that looks like a man.

Strange rain

Does it make sense if it's raining spiders, fish, frogs? ..

Actually, yes ... In Honduras .. This is a small state that is located in Central America .. In the provinces € exactly .. This phenomenon is called fish rain, and it has occupied scientists for a century. This phenomenon occurs in May and July of every year, when a large black cloud with thunder and lightning forms, and then large numbers of fish fall.

These fish are also edible, but scientists have found that these fish do not have eyes.

Okay, where do these fish come from? ..
This question puzzled scientists and stunned them ..
Of course, most of the theories developed to explain the fish rain phenomenon boil down to the fact that the source of fish is the ocean, and that hurricanes carry this fish with them and throw it on land with the rain.

But this theory seems far-fetched when it comes to rainfish in Honduras ... Why? ...

Since the city of the euro, where this fish ends up annually, is about 140 miles from the ocean, hurricanes do not make sense to carry all of this fish so long distance to throw them in this remote city than in the rest of Honduras.

So where does the fish come from? ..

The locals have a legend in euros .. He claims that one of the Saints visited their city a long time ago, and that he saw extreme poverty, hunger and poverty of the population, and he prayed to God for 3 days and 3 nights, and after that time came a big black cloud and spilled a lot of fish on the city, so the poor took this fish and ate it.
A beautiful legend .. But this is unconvincing for scientists who do not believe in occultism .. Therefore, after a long search, they came up with a new theory that the source of these fish is not the ocean, but rivers and springs near the city of Yoru , where they are carried by hurricanes and strong winds from there and are thrown on the city.

Of course, there is no evidence confirming the validity of this theory, but its systemists remain the most inclined .. The truth about the origin of these wonderful fish remains that only Allah Almighty knows.

Of course, fish showers have happened and are happening in other countries of the world, they fell in Singapore, India, Australia, America and the Philippines .. In another place, but not at the same pace and oddity as in the Honduran city of Euro.

Ibn al- Jawzi, in his book "The Regular", similarly reports accidents in 428 AD. e. :

"And in the spring another: a book came out of the mouth of reconciliation, which said that the people of the people of the mountain responded and it was said that they shed a lot of rain, during which the thickness of some of them weighed a pound and two pounds."

Showers not only bring fish with them, they can also carry more exotic things ..
In the city of San Antonio da Belti, in Brazil, we planted spiders from the sky instead of rain! .. And these spiders were hitting the power wires and everything else and trying to stick to it so the wind wouldn't wash them away.

Scientists interpreted this phenomenon as the result of a powerful storm that carried spiders with it and transported them to another remote location. The second explanation is that the huge web of spiders caught between the two pillars broke off and caused the spiders to fall like rain.

Are frogs falling from the sky too? ... Just like in the famous clip from the movie Magnolia (Magnolia - 1999) ..

There are many historical documents that show us that the sky rained frogs in peony and darvania , and frogs filled all the streets, and people took them to kill them,

but to no avail, because there were so many of them and ruined the water, food and crops. the inhabitants of the two cities found no other solution but to leave and leave their country.
In the book about the animal, he said: "Harith claimed that he was anyway if the blood cloud was so thick that it almost touched the ground and almost touched the tops of their heads, and that they heard both the sounds of magakakak and the roar in them. Stallions at Ashval , and then they were pushed by the strongest rain seen or heard until they succumbed to drowning, and then they fell.

In such cases, in 1944, frogs hit the Soviet Union, India and Australia, as well as England. In Bulgaria in 2005, a rain of small frogs fell from the sky, which one eyewitness described as: "I saw countless small frogs falling from the sky."

Other strange creatures appeared in the sky: jellyfish in England in 1894 and worms in Louisiana in the United States in 2007. In 2014, tens of thousands of bats fell dead from the sky in South Australia .. She fell dead due to unprecedented high temperatures in the area, and the smell of rot on the streets caused great harm to residents
Gifts from heaven can include things you don't plan on using, like one of the fishing boats in the Sea of Japan wrecked and drowned because a cow fell from the sky! .. .. This question puzzled everyone .. Later it turned out that a Russian plane was transporting a cow over the Sea of Japan when the crew decided to get rid of the cow at an altitude of a thousand feet, because she panicked and acted violently, which threatened to crash the plane, and by a strange coincidence circumstances wants this cow to land on a Japanese fishing boat! ...

In 2014, people were surprised by the fall of a human body on a street in Jeddah , Saudi Arabia, the body of a dark-skinned man with a severe head injury, apparently falling from an airplane.

In 1876, in the American state of Kentucky, pieces and pieces of meat fell from the sky, and the Press wrote about this for a long time, which puzzled people, and later laboratory research found that the pieces were horse meat, and the secret of their falling from the sky remained a mystery to this day. especially if we know there was no plane.

Blood rain poured from the sky .. Over the centuries, historical texts have told us that the sky has poured blood rain several times, and people often viewed the blood red rain as a sign of the anger of heaven, and its fall was combined with the onset of a big event on Earth, and this was considered bad omen or omen of the hour.
Scientists have often attributed the cause of the bloody rain to the red sand and soil brought by the wind from the deserts and wastelands, mixing with water and giving the rain its characteristic red color. But the bloody rains that hit Kerala in India since 2001 forced scientists to change their minds about the source of the bloody rain, water samples collected and analyzed in laboratories showed the presence of mysterious strange red particles that give the rain that characteristic red color, these molecules are alive , but they lack DNA, which increased the bewilderment of scientists, their opinions are different.
Blood rain is a source of terror for people, I don't think anyone would want their clothes to be stained with blood falling from the sky .. But there are other things that people

desperately dream of .. Poor people often raised their hands to the sky asking God to shower him with money and gold! ... On rare occasions these wishes have come true .. There are many stories of coins falling in the rain, sometimes there are treasures buried in deserts and remote places, carried away by hurricanes and dumped into populated places.

But in most cases, when money fell from the sky, it was not caused by hurricanes, but by people. Perhaps the most characteristic feature of money rain is the kind of people who are trying to collect the Securities scattered in the air, and the pure happiness of making this money without problems and fatigue .. Perhaps for this reason, that is, to make people happy, a rich American chose an innovative a way to say goodbye to this life and make people wholeheartedly take pity on his soul by recommending that ten thousand dollars of his fortune be thrown by a helicopter over the city where he lived, and as soon as this amount was thrown, so today anyone in this city who you ask about this deceased richer reminds him of love and pride, perhaps more than his father! ...
Another funny incident occurred in the suburbs of the American city of Miami one morning in 2010, on that happy day for the residents of this area, on one of the bridges that cut through their area, a strange incident occurred where an armored truck carrying a large amount of money collided and overturned and the truck crashed, the money it was transporting was scattered and ".. Crowds of people flocked to see the shares descending from the sky, but some did not bother to escape as the money landed in the gardens of their houses or entered the windows of their rooms .. What a good morning ..

People collected about half a million dollars in securities that day.

The funniest thing in this story is that police cars began to ride through the streets and ask residents to return the collected money through the loudspeakers .. But no one came to the police station to give what he collected, except the old woman and the boy .. The old man came for 19 دولار. And the boy came for 85 cents .. Less than one dollar!

Neutrino .. Phantom particle

In the midst of the massive discovery of the elementary particles that make up the atom (there are now over 130 species), this tiny particle occupies the throne of all of them in terms of mystery and excitement .. In ancient Arabian Wisdom, it was believed that "a thing whenever it hits violence" since it is one of the smallest particles ever - if not the smallest actually - scientists knew little about it and missed a lot due to the difficulty of capturing it, because if you want to measure it, it must collide with the nucleus of an atom and fall.

The reason it is hidden from scientists is because it has the following qualities:
It may have mass like the rest of the particles, but it has no charge, making it difficult to hunt because in this miracle, it moves around the universe at the speed of light - and can exceed it - and does not worry about any gravity as long as the black hole with the highest gravity known today, this particle will not pass by and disappear.

He can also pass between different parallel worlds without too much trouble, hitting classical physics against a wall as he walks and returns before the photons of light arrive at the station to complete the journey, and in this way he tricked some scientists as they thought he had speed. greater than the speed of light, despite the impossibility of this.

To be able to repel a particle, you must put up a barrier made of a specific material, the thickness of which varies depending on the penetrating power and permeability of that particle, for example, to repel alpha particles, bring a paper barrier. gamma rays need at least an inch thick, like a lead barrier, to prevent it from advancing. Cosmic rays are highly permeable, they are highly penetrating, so you must put up a ten meter diameter lead barrier to prevent this.

But this little genie neutrino is not one of those weak in his eyes, since you need a hail of lead with a diameter of $3 * 10 ^ 21$ kilometers to keep it from moving to the periphery of the universe.
The derivation of neutrinos is due to the phenomenon of the decomposition of some radioactive isotopes due to the release of beta radiation, which is played by
electrons. When one radioactive element decomposes into another, there is a certain

energy loss, this energy loss is the difference between the energy of the radioactive element and the energy of the resulting element. In order to comply with the law of non-annihilation of energy, an electron - from the nucleus of an atom and outward in the form of a beam of beta rays - must carry this difference in energy, but measurements have shown that the electron carries less energy than the energy superimposed upon decomposition, so the American scientist Wolfgang Pauli in In 1930, he suggested the existence of a small particle that carries energy that we do not see, and called it "neutrino" because it does not carry an electric charge.

It took scientists a long time to discover the neutrino with its three types (neutrino-electron, neutrino-muon and neutrino town). The discoveries were made in stages, from the 1960s to the end of the 2000s. It is believed that about 50 trillion solar neutrinos enter the human body every second.

At the end of 2011, the Opera team of the CERN project announced statements from previous experiments conducted in the 1980s that the speed of muon neutrinos is slightly faster than the speed of light (1.00005 times the speed of light) and even with all the validity of the statement, this is still remained within the framework of the principle of suspicion. That could rewrite the laws of relativity and modern physics, but the experiments were returned after criticism of some other scientists showed that there were doubts about the conditions of the experiment.

One of the Egyptian scientists, who worked in correspondence with CERN, proved that he could not reach a speed exceeding the speed of light, if he did not move between worlds, where he secretly moved to a parallel world, and was replaced by another, his twin, and thus prepared the tunnel in order to reach its replacement in less time than if he himself continued to move.

In fact, a neutrino - this elusive ghostly particle - uses its three flavors or facets during its journey between the folds of the immense universe.as it moves from a neutrino-electron, for example, to a neutrino-city and so on. it has puzzled twenty-first century scientists as to what aspects it really is. but this is just a trick used by this rogue, he moves to other worlds and places his place in our double world in order to maintain the law of constancy of the amount of matter in the universe.

Parallel worlds .. Between the myths of the past and the realism of science in the present

Parallel worlds .. Between the myths of the past and the realism of science in the present (1)
Author: Dr. Mohammed Zuhair
Parallel worlds .. Between the myths of the past and the realism of science in the present (1)
Is there another world in which other copies of us ..

There is no doubt that many of us have heard or read about the tales of a thousand and one nights, those tales that were conceived by US Scheherazade in the most creative description of magical worlds filled with imagination, and among these wonderful tales is the story of a boy Mandu, who lived an unhappy life in poverty, loved the sultan's daughter and when two strange creatures forcibly took him to a distant planet at the edge of the galaxy and presented him to their king, claiming that he speaks their wisdom and may have come to their knowledge. The state of the king and a promise to help him by sending him to a world parallel to his world, where he will find another Mando who has already solved his problem, and later he will become king and learn from him how to solve his own problem.

A story that some of us may consider just a myth from mythology, which has no connection with reality, but is it really so?

Are we alone in this universe? ... Have you ever had this question before? !!

A question that has been repeated so often that it has become a common question, but the answer is often yes or no ... "the topic has not been decided yet" !!
The theory of the multiverse (the Multiverse):
Our telescopes are capable of seeing a maximum of less than 14 billion light-years ... But what next? From all that is known to scientists, we know that the universe arose as a result of the great bang ... But what happened before that? The questions that philosophers, religious people and thinkers of different eras have tried to answer: what is behind the universe that we see? And how did the universe come about? And where? Scientists believe that today we are able to give logical, balanced and far from the ideas that prevailed in the past.

Until recently, all scientists were engaged only in studying the Universe with its contents and its physical laws, as well as what happened after the great bang 13.7 billion years ago ... Now science is breaking into the area that was once the reserve of philosophy and religion: what happened before the birth of the universe? And what is beyond its known limits? ...

Is there something else?

Let's say there are universes or universes that are different from the one in which we live. :

Imagine that you have a version of you in more than one universe and more than one Earth ... "Omar is not Omar and Ali is not Ali you know ..." They are in their own universe , you are in one universe, you work as an engineer, but in another universe you work as a lawyer or maybe at a gas station or somewhere else !. Multiple universes, but you're more of a different version !!

Here's what the multiverse theories say : There is more than one "more than one" version. How did this theory come about?
In 1954, a doctoral student at Princeton University, Hugh Everitt, put forward a radical idea: there are parallel universes, just like ours. All these universes are connected with us, in fact they are ramified from us, and our universe is also ramified from others.

In these parallel universes, our wars end differently than what we know the extinct species in our universe have evolved and adapted to others. In other universes, we humans may have become extinct.

This thinking boggles the mind and is still difficult to understand. General ideas about universes or parallel dimensions, similar to us, appeared in the works of science fiction. But why would a young physicist with a future risk his career by presenting the theory of parallel universes? ...

In his theory of parallel universes, Everett tried to answer a difficult question in quantum physics: Why do quantum objects behave badly? .. The quantitative level is the smallest of all that science has discovered so far. The study of quantum physics began in 1900 when scientist Max Planck first presented this concept to the scientific community. Planck's research on radiation has prompted some discoveries that contradict the traditional laws of physics. These discoveries suggested that other laws exist in this universe, operating at deeper levels than those we know.
In the short term, physicists who have studied the quantum level have noticed strange things in this world. First, the molecules present at this level take arbitrarily different shapes. For example, scientists have noticed that photons (small beams of light) behave like particles and like waves! Even one photon makes this rotation at the same moment. Imagine being visible and acting like a dense human being when the friend looks at you, but when he turns to you again, you turn into gas! ...

This case was known as the Heisenberg Uncertainty Principle. Physicist Werner Heisenberg suggested that by observing quantum matter, We influence its behavior. Thus, we cannot be completely sure of either the nature of a quantum thing or its characteristic qualities, such as speed and location.

This idea was supported by the Copenhagen interpretation of quantum mechanics. This interpretation was put forward by the Danish physicist Niels Bohr. That quantum particles do not stop at one particular case or another, but in all these cases the potential is simultaneously. The sum of the states of a quantum thing is called the wave function, and the state of a thing existing in all its possible states at the same time is called superposition.

According to Bauer, when we observe something quantum, we influence its behavior. Observation violates the Supermode state of a thing and forces it to choose one state from its wave function. This theory explains why physicists get conflicting measurements from the same quantum thing: a quantum thing chooses different states during successive measurements.

As exciting as it sounds, Everett's multiverse theory has implications beyond the quantum level. If there is an event that has more than one possible outcome, then - if Evert's theory is correct - the universe will branch out when that event occurs. This actually happens even if the person decides not to do anything.
This means that if you are in a situation where death is a possible outcome, then in a parallel universe you are dead. This is just one of the reasons why the theory of multiple worlds upsets some.

Another troubling aspect of interpreting parallel worlds is that it destroys our linear concept of time. Imagine the timeline depicting the history of the Vietnam War. Instead of a straight timeline showing the progress of noteworthy events, the timeline in multiverse theory branched out to show all the possible outcomes of each event. For example, you met your future life partner in your world, and then got married and gave birth to children after a while, but in another world you can break up and the marriage will not take place, so every possible outcome of the event will be dated.

But a person cannot know about his other twins - or even about his own death - who are in parallel universes. Therefore, it is difficult to predict that one of the universes affects the other parallel or even to test it.

But the multiple worlds theory isn't the only theory that wants to explain the universe. And she is also not the only one who offers us parallel universes.

String theory
Since their science has progressed, physicists have been busy re-engineering the universe, studying what they could observe, and working from behind on smaller and smaller levels of the physical world. With this work, physicists are trying to reach the ultimate and most elementary level. This is the level that they hope will help establish an understanding of everything else.

After his famous theory of relativity, Albert Einstein spent the rest of his life searching for the final level that would answer all physical questions. Physicists call this ghost theory Theory of Everything, which quantum physicists believe is on the way to discovering this ultimate theory. But another area of physics believes that the quantum level is not the smallest level, so it cannot give us a theory of everything.

Instead, these physicists have turned to a subquantum theoretical level called String Theory to answer all of life's questions.

Surprisingly, these physicists, in the course of their theoretical studies, also deduced - like Evert - the existence of parallel universes!

String theory was developed by the Japanese-American physicist Michio Kaku . His theory states that the basic building blocks of all materials, as well as all physical forces in the universe - such as gravity - exist in the subquantum plane. These building blocks are like very small rubber bands - or strings - that make up quarks (quantum particles), electrons, atoms, cells, and so on. Exactly what kind of material is produced by the strings, as well as the behavior of that material, is determined by the vibrations of those strings. The strings vibrate, generating various forces that govern the universe. Thus, our entire universe is music. According to string theory, this game takes place in 11 separate dimensions.

Like multiverse theory, string theory shows the existence of parallel universes. These dimensions are so wrapped around ourselves that we are not aware of them in our world. Since the universe has these different dimensions with its diverse geometries, and knowing that the laws of nature depend on the geometry of nature, it is expected that these multiple dimensions will shape the different universes in their laws and realities. This is how string theory leads to the conclusion that there are many different universes. In addition, Kaku refers to the fact that billions of solutions to string theory equations have been discovered, and each of these solutions describes a mathematically symmetric universe and differs from other universes described by other solutions of the theory. Thus, string theory assumes the existence of several universes.

According to the theory, our universe is like a bubble next to similar parallel universes. Unlike multiple worlds theory, string theory suggests that these universes can be in contact with each other. String theory states that gravity can flow between these parallel universes. And when these universes interact, there is a big bang similar to the one that created our universe.

*There are some theoretical explanations for some of the experiments that prove that there are at least 8 × 6710 parallel universes, different from our universe and our world, which is close to and even mixed with us in the same space in which we live, including a number of moments time, which create a parallel world, representing the number 2 * 4010, which are worlds in which each of us is present and in the past .*

... At CERN, I had the assumption that if we want to move from our world to another parallel world sharing the same space, the key to this lies in gravity, because if we study gravitons (very small particles in an atom that carry the forces of gravity), we will certainly come up with a way to use the forces of these particles to go to other parallel worlds. Let me offer a scientific explanation (after reviewing it with some scholars of Al- Azhar) two sacred verses 14,15 Sura al- Hajar , in which I find an accurate description of this hypothesis and their text: "If we opened the door for them from heaven and remained lame in her (14), they would say our eyes were closed.)

Let's take a look at the words and walk together. :

Truth tells us that if unbelievers were destined during the reign of the Messenger of Allah (s) to open by them (that is from above) by the power of Allah Almighty the door from heaven (that is, the passage that was blocked and closed when the door that is from the structure heaven and which is held by gravity), then that would be true. - Unless the Ascension process is influenced by the forces of gravity (which is used with this hypothesis when transitioning between worlds), look at the word " fahlwa ", which will show us that it will describe their emotions and reactions during the transition process in the next verse. and their reaction was to say, "but our eyes are drunk, but we are enchanted people," because they see Western things happening around them from the impulse only it is more like magic that impresses the human soul with extreme awe. And Almighty God is above, and I know.

Goodbye, God willing, to the second episode of the parallel worlds series, in which we get to the idea of cosmic worlds throughout history and how many cultures it has.

The Tamam Shad case

Jeweler John Lyons was on a trip with his wife to Summerton Beach at seven o'clock on the evening of November 30, 1948, twenty yards away was a neat man sitting reclined on the wall of the sea by the shore, legs stretched out and legs crossed with each other. noticed him stretching his right arm out as far as he could and then letting him fall casually as if he was drunk.

Half an hour later Mrs. Olive Neal and her companion, Mr. Jourdain, looked out from behind the wall and saw the same man lying in the same place while the street lamps were on. The lady said there were other people there, including a man in his fifties who stood and looked in a certain direction, and she thought that Mr. Neil was looking at this man. According to them, he was lying in the same place and never moved in the half hour that they saw him, and mosquitoes circled around his face, and he did not try to restrain him, and Mrs. Neal drew the attention of her companion, asking if he was asleep. he, and he looked at him for a while and sarcastically expected him to be dead.,
The next day, Mr. Lyons went at 6.50am to go for a swim on the same beach that he and his wife were hiking last night, and when he got out of the water he found his friend and they started chatting .vo their conversation, they noticed that a group of people gathering around the dead man, who was lying in the same spot where he saw a stranger in the night. when Mr. Lyons approached and saw him, he suspected that he was the same person, so he went to his house and reported it to the police.

Brighton Police sent Constable Moss to the scene to learn about what would later become the most mysterious and bizarre crime in Australian history, with no victim, no killer, and no cause of death. A crime full of mysteries, mysteries and strange events associated with strange people and associated with other crimes no less strange.
Mysterious dead
He was lying on the shore by the sea wall exactly where the witnesses saw him last night, his legs were stiff and pointed towards the sea, his left arm was straight and his right arm was bent, his organs began to numb, and he did not seem to be subjected to any violence , whether it be a knife, He looked like a dead quiet peace and threw a cigarette before lighting a cigarette and signed on the lapel of his jacket. it was correct that he pressed on her chin to hold her in place, but nevertheless did not disturb the consequences of the burns, and also discovered another cigarette, not smoked after being placed behind her ear.

The man was wearing a white shirt with a red tie with blue and white stripes, a brown pullover with a coat over it and brown trousers, and in one of his pockets he had an orange thread that was not suitable for the procedure, suggesting that he might , did it on an emergency basis. He also wore two socks and shiny lace-up boots that didn't have a hat on, which was weird at the time. His clothes were suggestive of style, but wearing these clothes on top of each other in such a hot atmosphere was surprising, as he dressed in a hurry, not thinking about what he was wearing, or maybe he had nowhere to put his clothes Each other.

Something surprising was noticed on his clothes, a sticky card containing information about the dress, which is usually a piece of fabric sewn somewhere inside the dress, all these cards were removed from the clothes of the deceased, as if someone was trying to hide any information , which could lead to the identification of this person.

Following a field investigation, the mysterious body was taken to the Royal Adelaide Hospital, and as a routine procedure, Dr. John Barkley formally declared the man dead, justified the cause of death with cardiac arrest, and poison was likely the cause, and set the time of death at 2 a.m., and then transported the body to the morgue.

Person without personality
While the doctors were doing the autopsy, it was extremely difficult for the investigators to identify the victim, since he did not have any identity documents. They looked into all the reported missing persons and found nothing that matched the victim's description.

The investigators resorted to a solution they never liked and did not want to resort to - to seek help from the press to identify the murdered person. Police usually avoid this remedy because anyone with a missing relative will think the body belongs to him and there will be thousands of people claiming to know him, forcing investigators to check each statement individually, adding to the effort and making it difficult to focus on the case.

In its December 2 issue, The Observer , on page three, reported that it was suspected that the body belonged to Mr. EK Johnson, 45, but the next day a man walked into the police station and told them that he was Mr. EK. Johnson, who is alive and well, and thus brought them back to ground zero.

The Evening News on the same day posted a photo of the victim on its front page, and calls to the police station were flooded with people claiming to know the owner of the body.

Two people went to the police station to confirm that they identified the victim as Mr. Robert Walsh , 63 , a lumberjack, and Mr. Walsh went to buy sheep in Queensland a few months ago , but never returned.

A man named James Mack contacted the police to identify the victim, but when he looked at the body he could not, and an hour after leaving the police station, he called them and said that when he saw the body, he was not sure at first because of the victim's hair color, but now he's sure it was Mr. Robert Walsh .

But Miss Elizabeth Thompson, one of the two people who initially allegedly identified the victim as Mr. Walsh when she saw him a second time, changed her mind and said it wasn't him.

Police initially was not sure that this is Mr. Walsh , and the statements of Elizabeth Thompson reinforced her suspicions: Mr. Walsh was 63 years old, while the age of the victim does not exceed 45 let.krome addition, Mr. Walsh said to have engaged in logging, and the effect of was evident on his hands, but the victim's hands were polite. if he has been logging, he must not have been in the past 18 months, which does not fit Mr. Walsh 's description .

In the months that followed, police frequently made statements regarding the deceased's identity, until in 1953 there were 251 statements in which the owners claimed to have seen the deceased or knew who he was. But none of these statements were helpful to the company.

Investigators tried to identify him by fingerprints, and all of his fingers were imprinted and matched with all fingerprints in Australia and then in all English-speaking countries, and they turned to Scotland Yard and the American FBI for help, but without much result .. How as if this dead man had never existed on this earth! ..

Autopsy
In the morgue, the man was stripped and numbered, his pockets turned out and his belongings laid out on the table. He had a second-class train ticket from town to the coastal suburb of Henley Beach, which was not yet in use, and a bus ticket to town, a bus stop a thousand hundred meters north of the beach where he died.

In addition, he had an American-style metal hairbrush, a pack of Armyclub cigarettes , Bryant and May matches , chewing gum and candy. But he did not have any identity documents, no passport, no money, not even a wallet, maybe he had money, but they were stolen along with the wallet, no one knows.

An autopsy by pathologist Dr. John Dwyer revealed that the victim's last meal was a meat pie four hours before death and was mixed in his stomach with a lot of blood, indicating that he was poisoned, but analysis of the remains of the pie showed poisoning was not from her.

Perhaps this was the reason for the unusual behavior of the man when he was seen that night on the beach in clothes that did not match the atmosphere of the place, and the way he stretched out his right hand and then casually lowered it, causing bystanders to believe that he was drunk ...

His internal organs were in an abnormal state when his brain was stagnant, as well as his stomach, intestines, kidneys and liver, and his spleen was three times its normal size, which caused anxiety, and the doctor noticed that the blood vessels of the liver were blood flow. These were clear signs of poisoning, and it was suspected that he had committed suicide by ingesting poison.

A chemical expert re-examined the blood and internal organs at the request of
Dr. Dwyer to trace the poison and find out its source, but the result was shocking and
surprising, there was no trace of any poison in the organs and blood of the victim. "I am
very surprised that he did not find anything, but I am absolutely certain that the death
was not caused by natural causes, but I am convinced that it was caused by a poison that
may have been swallowed in a sleeping pill," Dr. Dwyer said in his statement.
Brown bag
On January 12, 1949, the case took a strange turn. Since the man had a ticket on which he
came to the city, the police expected that the man had a bag of clothes with him and that
he could put it somewhere, all hotels and hotels in the city of Adelaide were notified, and
the effort paid off, so how the Adelaide train station staff found a brown leather suitcase
that was also packed.

The bag had a red square nightgown with a pattern, several pajamas, red felt slippers,
four pairs of underwear, four pairs of socks, a shaving set with mousse, a cooler and a
shaving brush, our trousers were light brown and had a little sand in them , a sewing kit
with an orange spool of thread, two ties, three pencils, scissors and a knife sharpened for
a weapon, a tin box, an electric screwdriver, and a painting brush, commonly used by a
naval officer on cargo ships to train commercial cargo.

The contents showed that the bag was prepared for a stay of no more than two days or a
maximum of a week, but there was only one shirt, if this bag belonged to the deceased,
then he did not look like a person wearing only one shirt for a week.

The police were able to tie the bag to the deceased because the spool of thread was
exactly the same type and color as the one that the deceased used to mend his coat pocket
when he died on the beach, and the thread was not the only proof that it was his bag.

After seeing the clothes in the bag, one of the experts confirmed that the sewing used in
processing the clothes was from the manufacture of sewing machines found only in
America, and that these clothes could not be imported but brought with him when he
came to Australia from America. or bought it from someone who came from there.

One of the important things about this bag is the quality of the sticker on the tie with the
name T Keahe and another on the jacket with the name Kean , as well as on the laundry
bag with the name Keane . Extensive research by maritime agencies showed that none of
them bore the name Ken or Kenny at all.

But why were all cards removed from the victim's clothes except for these?!.

The police believe that she was deliberately left unattended because the deceased's name
is not Ken.

These outfits (jacket and tie) we had a special combination of numbers used in laundries,
these numbers were applied to the clothes upon receipt, in order to be replaced in case of

loss of the receipt. The police conducted a thorough search of all laundries in the country to find a laundry that used this combination of numbers but never found it.

Further investigation
Four months after the discovery of the mysterious man's body, the police asked the pathologist and professor at the University of Adelaide, Mr. John Cleland, to re-examine the body. Cleland explained after examination that the body belonged to a man between the ages of forty and forty-five, he was about 180 cm tall, broad-shouldered, with a slender middle, that is, he had an athletic build, his eyes were honey-colored, and his hair was blonde with some gray and it was clear from his hands that he was dead.

Mr. Cleland also noted that the toes were pointed as if their wearer was wearing a tight shoe - perhaps as a dancer or as someone who had to wear this type of shoe to be beautiful. His leg muscles were as bumpy and taut as those of long-distance runners, ballerinas or cyclists, his body was surprisingly athletic for a man of his age, and his legs were tanned as if he were wearing shorts, perhaps due to the working conditions.

Re-examination by Dr. John Cleland revealed several other things: the victim's shoes that he wore at the time of death were clean and freshly polished , which is not the case for the shoes of the man who allegedly wandered around the city before he came to the beach. indicates that he died somewhere else and was transported to where he was found, especially since this corresponds to the absence of vomiting or vomiting.

One investigator who recently investigated the case said he found testimony in police records given by a person during the investigation when he said that on the night of the incident on the beach he saw a person carrying another person on his shoulder at the water's edge and heading for the sea wall, but he did not see his face, But the common denominator in the testimony of all the witnesses is that he was not one of them who saw the face of a man who was sitting at the sea wall that night, and none of them are sure what it is was the same one, was found dead the next day.

Quatrains about tents and Tammam Shad clip
Dr. Cleland's find added even more mystery and strangeness to the case; a search of the deceased's clothes revealed a small secret pocket sewn into the part of the trouser belt that the former specialist who examined the body did not notice, inside this small pocket was found folded paper on which it was printed following sentence (TAMAM SHUD), this paper was taken from the pocket of the deceased.

The local library staff were called in to translate the sentence and reported that it was a Persian word for "finished" and was written in Persian by Tamam Shad . The police used all their energy to unearth the book, from which this copy was removed in all libraries and bookstores across the country, a photograph of the copy was published in all state police departments and released to the general public.
Weird coincidence.
On July 22, eight months after the incident as the police continued to publicize the topic, including the clip image, a strange coincidence occurred: a man named Ronald Francis went for a walk with his brother-in-law Hellman Minx in the latter's car,

which he parked for several days without locking her. But after the newspapers reported on the clip and saw it, he called his son-in-law to find out where he got the book he saw in his car. Tell him that he found the book in the back seat of his car when he parked on Mussolini Street, but he doesn't know who put it there and prevented his property from being returned.

Mr. Francis took the book, took it to the police station and handed it to them after he told the whole story, and when the paper was compared to the book, it immediately turned out that it was removed from this book, which is strange, since this edition of the book is the first since since it was printed in 1859 through Whitcomb and Tumbes and is therefore considered valuable and rare ..

The oddities did not end there, on the last page of the book, the police found a blade written in pencil, but it is not known who the author was, whether he was the victim or someone else, the blade was written in five lines in a row, and it was clear that the second line was crossed out and it was not:

WRGOABABD MALIAOI WTBIMPANETP MALIABOAIAQC ITTMTSAMSTGAB

It was not clear from the handwriting whether the first two letters of the first two words in the first and second lines are M or W, but it is believed to be W because it is different from writing the letter M in the fourth line. And above the letter "O" in the fourth line was the letter "X". The police weren't sure if the mark made sense in the code or if it was accidentally placed, the police weren't sure if they were in front of the code or just meaningless graffiti. This is why cryptographers in Australia were contacted and called in to decipher this code, but failed.

When the code was analyzed by the Australian Department of Defense, they said the codes were incomplete, making it impossible to provide a sample to work with, or that they could be complex alternate code and possibly meaningless answers coming from a restless mind. Let them know that it is impossible to give satisfactory answers.

Donated to the University of Adelaide to decipher this code, but it also did not pay off, they needed to see the first edition of Omar Khayyam's Rubayyat book , but they did not have a single copy of this edition, not even the women who brought Ronald Francis to the police were police officers and why got rid of them.

Unknown nurse
This code or scrawl emanating from a restless mind was not the only thing found on the last page of the book, there was a telephone number belonging to a former nurse who lived on Moseley Street, this nurse named Teresa Paul or Teresa Johnson, who lived on Moseley Street just above Summerton Beach, where the victim died.

The police went to Teresa's house to check on her, telling the police that she was not at home that evening, that is, November 30, but one of her neighbors told her that he heard a stranger call her at home the same night she was not at home. The police, of course, did not know who the man was.

When asked about the phone number in a book of poems found in Hillman Meeks' car, she replied that she had a copy of the book when she was a nurse at the Royal North Shore Hospital in 1945 and gave it to a soldier named Alfred Boxell who was said to have worked in an intelligence unit. But that didn't answer the question of why her phone number ended up on the last page of the book.

Teresa was shown a template taken from the face of the unidentified body, and the police officer who showed it to her revealed that she was very shocked and almost passed out when she saw it. She is a nurse and used to seeing the dead. what scared her? .- Did she know who he was? ... Judging by her reaction, she knew him, but argued otherwise.

The police were looking for this Alfred Boxel , hoping to find something from him that could solve the mystery of the case, and indeed found him alive and working in a bus repair in Randwick . He told them that he did not know who was killed, but he said something that added to the mystery and complexity of the case. She rewrote Quartet No. 70 on the first page of the book. :

True, true, I repented a lot before I swore, but were you sober when I swore, then spring came with flowers by hand, my old remorse was torn to pieces

Teresa pleaded with the police not to disclose her name in the investigation because she is a married woman and does not want anything to harm her reputation and affect her marriage, especially since her husband does not know anything about it and for her the past is over. And I asked them to name her in my investigation and in the materials of the Justine case , under this name was signed a copy of the quatrains of the tents, which I gave to Alfred. By the obvious stupidity of the police, they agreed to her request and did not investigate it further, and if they did, they would find many leads, the first of which is the identity of the murdered person, because Teresa knows exactly who he is. But they preferred to keep her real name a secret, which was revealed only later by mistake.

You didn't know Teresa was not married when I told them so and begged them not to ruin her marriage. Teresa was a suspicious and curious person when she worked in the hospital in 1945, she was known as Justine and she was not married, then moved to her mother's house in Melbourne and gave birth to a child with an unknown father, then moved to Adelaide, and the name Johnson gave the police was that of her lover.

Retired detective Gerald Feltz interviewed Teresa in 2002 for his book on the case, which he titled "The Unknown Man," and called her evasive and said she had no desire to make any statements regarding the book of quatrains of tents or Alfred Boxel or the dead, stating that she knew nothing then or now. But he thinks she knows a lot.

After her husband's death, there was a hope that Teresa would reveal the secret she was so afraid of, but, contrary to expectations, she left in 2007, taking her secret with her to another world.

That the case is still open

The AIF took over the burial procedure, and a charity paid for the funeral and wrote on his tombstone (Here is an unknown person). It was said that a few years after his funeral, an unknown woman was seen at his grave, and once, when she left the cemetery, the police were waiting for her, they asked her if she knew the person she was visiting, but she denied this. Around the same time, a secretary working at the starthmoor hotel across from Adelaide train station contacted the police and told them about the incident, she said the stranger had booked room 21 at the hotel at the same time the body was found and left the hotel on November 30th, and when the janitor went to clean his room, she found a black medical bag.

Attempts to solve the problem continued in vain, many astronomers and mathematicians tried to decipher the code, but failure was always their ally. Detective Felts believes the code was the initials of words meaning a meaningful sentence, for example the last line is a sentence: (it's time to live in northern Australia on Moseley Street).

Tamam Shad's case is open to this day. Over the past decades, dozens of theories and hypotheses have appeared that have tried to solve this insoluble riddle. They said that this case has something to do with espionage, and they said that it was due to emotional caprice and adultery .. There are plenty of talk and rumors ... But no one could explain why the identity of the murdered person remains unknown, how is it possible that no one on the planet can identify this person and prove his identity, although hundreds of newspapers around the world have published a photo of him over the years, and there have been books and documentaries about him in different languages ..

Who is he?. What's his story? ... What is the mystery behind this? ... The questions puzzled and perplexed researchers .. It is not surprising, therefore, that the case of " Tamam Shad " with its strange clothes and mysterious details is one of the most famous and complex riddles and mysteries of the twentieth century.

Are they related?
In June 1949, in the mountains around Largs Bay , twenty kilometers north of Summerton , a sack was found with the body of a two-year-old boy named Cliff Magnussen , and his father Keith Magnussen was unconscious and fighting for his life. The man who found them, a man named Neil McRae , Neil said he had a dream about Keith and his son and the place where they were found, that is, the hills around Largs Bay , one night before he got them. found.

Kate and his young son disappeared four days ago, and the baby died the day before he was found, the pathologist could not determine the cause of death, but assumed it was poison, like that strange man on Summerton Beach .

When the police arrived at Magnussen's house to meet Keith's wife, she told them that something strange had happened to her after the accident with her son and husband, she said that a strange man who covered his face with a mask and drove an ice cream car , tried to run over her with his car in front of her house, then stopped the car and threatened neighbors, later told her that he saw a stranger watching her house and circling around, and she expected it to be the same person who threatened her and wanted to run over her with her car.

Ms. Magnussen believes that all this happened because her husband knew who the stranger was and wanted to inform the police that he suspected someone he worked with in 1939, Karl Thompson. Surprisingly, Miss Magnussen fainted shortly after being questioned by the police, demanding medical attention.

Mr. Goyer , secretary general of the Northern Union of Largs , received a phone call from an unknown person who threatened that Mrs Magnussen would be a terrible disaster if he tried to stop her in any way. Mr. Curtis , the mayor of Adelaide, also received three calls in which he received death threats if he poked his nose into this Magnussen family case .

Again the quatrains of tents
In Sydney in 1945, a young Singaporean named Joseph Saul , 34, was found dead in a bathtub, face down, next to an open book of tent quartets, and is believed to have died also poisoned, just two months before anonymous nurse Teresa presented a copy of the tent quartets to Alfred Boxel . Was it all a coincidence, or did the nurse play a strange role in these deaths? ...

This Joseph was the brother of the Prime Minister of Singapore, so a high-level investigation was launched to find out the circumstances of his death, and he had a hairdresser friend named Gwyneth Graham , who, shortly before his death, gave her money to start her own business as a hairdresser. , and later became a key witness to his murder, she was interrogated and released thirteen days later.

Time Travelers

Time travel is one of the topics that has been discussed, and many scientists still believe that this can happen, but the human mind has not yet reached the technology that allows it to travel back and forth in time to the past and the future.

Some believe that the story of the Companions of the Cave is a people who traveled in time, where they woke up three hundred years later to find themselves in another time, and the people who were with them left and left after some other religion, which turned to the priests for its distribution.
HG Wells' novel The Time Machine was one of the most important works of fiction, which accurately depicts the idea that the protagonist travels into the future to see the events that will occur as a result of the present.

Scientists say that travel to the future can take the form of deep sleep, or stay frozen for 100 years and then wake up in another century, but travel to the past happens in the form of one and starts at speeds faster than the speed of light.

Does he see the future?
In 1935, British Air Force pilot Victor Goddard had a terrible experience: while flying from Edinburgh, Scotland, back to his base in Andover , England, he decided to fly his plane over an abandoned airfield in his sleep, near Edinburgh. Suddenly Goddard faced a storm accompanied by yellow and brown clouds, with which he lost control of his plane and began to descend in a vortex to the ground, but Goddard barely managed to avoid hitting the ground, he noticed that he was directing his plane against his will towards the abandoned dream airport, and as he approached it, the storm suddenly disappeared as 1935 arrived - the year of the accident.And there were the mechanics working at the airport wearing blue work uniforms, which Goddard found very odd, mechanics in the British Air Force usually wearing brown uniforms, and even weirder was that none of the mechanics working at the airport ever noticed him flying over them, suddenly the same storm came, but managed to avoid it and go to Andover .
The mystery behind this is that in 1939 the British Air Force changed the color of their aircraft to yellow, included single-deck aircraft in their crew and changed the mechanics' uniforms to blue. So, did Goddard fly four years into the future, where he saw what would happen in 1939 and then back to 1935? ...
Swiss watch in a Chinese temple
In 2008, Chinese archaeologists unearthed a 400-year-old indoor mausoleum that belonged to the Ming royal family, who ruled from 1368 to 1644. When they removed the dirt from the shroud in the mausoleum, a small piece of stone fell to the ground, but it made a sound similar to the sound of metal that attracted attention of those present. when they picked it up and found it to be a small ring, they removed the remaining dirt stuck in it and examined it carefully. From the shape of the ring it became clear that this was a small watch of a Swiss watch brand, and the hands showed ten and six minutes. This clock was 100 years old, and scholars are confident that they were the first to enter this tomb since the burial was placed in it for any 400 years, please note that at the time when the Ming dynasty ruled China, the clock in all its forms had not yet

appeared, and there were two journalists accompanying the Chinese scientists and documenting the discovery in the film.

Rudolf Wentz

In 1950, a strange man appeared in Times Square in New York, dressed in a torn Victorian costume and with two old-fashioned salwins on either side of his face, looking surprised and dazed and surprising the people who were there, with his figure, walking without a guide, and a racing car appeared and hit him and he just died.

In the morgue, the police found him holding a five-cent beer bottle with the name of an unknown saloon, and none of the old-timers knew of his existence, as well as a seventy-dollar coin from the nineteenth century, a bill for caring for a horse and cleaning a cart from one of the stables on Lexington Avenue , on Fifth Avenue, dated June 1876.

Detective Hubert Rhyme of the NYPD 's Missing Persons Department stepped in to verify the identity of this person using the information he found and went to the address shown on the business card as well as on the letter envelope, but found a company there that had nothing about did not know him. His name did not appear in any of his notebooks, he had no fingerprints, and he did not find anyone to report his disappearance.

He found in one of the address books of 1939 a mention of a man named Rudolf Wentz Jr., so he was delighted and went to the indicated address and spoke to the residents of the building, who told him that he was a 60-year-old man who worked nearby and after retiring in 1940 went to a place they didn't know about.

The detective went to the bank to inquire about him and was told that he died five years ago, but his widow is still alive and lives in Florida, when he called her he told her that her father-in-law already lives where he is mentioned on a business card found along with this man's belongings and she told them that when he was twenty-nine he already lived there.

It was said that this story is fiction, created by one of the authors of stories in one of the magazines, but one of the researchers in the news archive in Berlin found the News in April 1951 about the story, told as it is mentioned today, and that in exactly five months before the science fiction writer wrote his story, in addition, a number of researchers reported that they had found evidence of Adolf Hitler.

Traveler without a country

In April 1954, one of the planes arriving from a European country stopped at Haneda International Airport in Tokyo and dropped all of its passengers, including a tastefully dressed, middle-aged Caucasian who told customs officials that this was his third visit to Japan. His primary language was French, but he spoke several other languages such as Japanese. when they asked him about his country of origin, he told them that he came from a country called Taured . He insisted on this and showed them his passport, which had already been issued from a country called Tord , which does not exist on Earth, and the passport had visa stamps confirming his statements that he had already come to Japan and traveled to several others. European countries.

Asking the officials to mark his country on the map, he pointed his finger at the Principality of Andorra, a very small state located in southwestern Europe, the man was even more puzzled and angry, since he had never heard of the Principality of Andorra and *was surprised that his country did not exist on a map that has existed for almost a thousand years.*

Customs officers found several coins from various European countries on him, then took him to a local hotel and placed him in one of the rooms, accompanied by two people, to guard him until a solution to this puzzle was found.
The company he claimed to work for was vetted, but company officials made it clear that they did not know who he was and had never heard of him before. The hotel, which informed them that he had booked a room in it, also said it knew nothing about him, as well as the Japanese company he allegedly came with to do some work, but the Japanese company also denied knowing him in general, despite the presence of a large number of documents confirming his statement.

When the hotel room in which the person was placed was opened, the surprise was that the person was not present despite the presence of two guards at the door all the time and despite the fact that the room is located on a high floor and there is no balcony in its window. from which one could escape, not only this, but also that.

Who is this guy? And where did he go or how did he go? Rather, where did it come from? No one knows, after that he was never seen again and this mysterious mystery was never solved.

Devil drug

During several months of my research on what drugs are and what types, I have found common ones such as marijuana, hashish, SD, speed drugs and heroin ... But I have never heard of a drug called the Devil's drug, which of course , is not its real name, but it is so named because it is like the devil in its effect on the mind .. Let's get to know him better ..

Overview
Like other types of drugs, the devil drug is extracted from plants, in particular from a plant called Mimosa, which is widespread in South and Central America, where the climate is favorable for its cultivation.

This plant is usually planted in large areas due to frequent drug users in these areas, and sometimes exported to Mexico, Cuba, where the flayed from the roots of roots , to produce what is known the mind of the devil, and its scientific name - "de MTV."

Property used in scientific experiments was sometimes cashed in by my science fiction "for no reason" and for other purposes. He is only discharged with a medical permit.

The drug is available in various forms, including what is drunk, and among them in the form of tablets. He paints on them figurines of animals, girls and cartoon characters ... In a way to seduce the consumer, not to mention the bright colors that color the shell of the grain.

Drug action

To prove the seriousness of the drug, some amateur in the humanities tried it, and the result was amazing. One pill starts working 20 seconds after taking it, and the effect is very strange. After the rapist woke up, he began to talk about a world that does not exist, he said, as if I entered the crypt of my mind, wandered between places tangled like threads, and noticed non-existent creatures in our world, like a peacock with four eyes, hands and vivid colors, cooler and better than usual. He highlighted an important point by saying that he met characters who have died long ago.

Other people secretly experimented with the drug, and as a result, one of them took it at home and found an hour later that he was in the tombs! ...

In another experiment by an amateur from California - a lover of the Paranormal and Gin - he says that when he took the drug, he felt the world around him exfoliate and transform into another form, and that he personally met the king of the genie and spent an hour with him. But how did he know he had spent so much time here? ... Simply because the drug works only for an hour, after which the user returns from his constant delirium to reality.

Science opinion about this property

Psychologists compare these cases to conscious delusions or hallucinations perceived by the senses, because the drug of secretions goes back to the brain cells directly and for this effect begins at a speed of 20 seconds, in the cases indicated at the top of the article he explained that the mind mixes fantasy and reality shows a vivid user picture

A person who saw the dead in his hallucinations, this is considered a state of approaching madness, perhaps he saw their spectra, because he always thinks about them, but it is impossible to see them in their original images, because the mind is in a state of dormancy or scattering, is not aware the images that he reflects to the offender, so he cannot see them.

As for the last case of the person who saw the king of the Jinn, this is the greatest evidence that in the hallucinatory state the mind displays images in memory in a random way, which we always think of.

For these reasons, governments impose strict control over it and imprison anyone who sells it without an international license because of its seriousness. Those who often consume it in the form of buns costing up to 440 are mostly in psychiatric hospitals. are treated for brain dysfunction between perception of reality and fantasy hallucinations.

Property in literature
Since I am one of the pioneers of general reading, Ahmed Murad 's Blue Elephant was the first to open the door for me to explore this strange subject, it is true that the novel carries some misconceptions about property, but I took advantage of that and the cover illustrates the situation I mentioned in relation to property.

Ride Into This Digital Dope

I haven't done extensive research on this yet, which made me feel insecure, especially after watching video sites , news sites and content that talks about it .. I was so overwhelmed .. Not because it happened .. This is no stranger human demons, but what amazed me was the availability of these ringtones on Youtube in English and Arabic.

This means that we are surrounded by a whole bunch of games ... we didn't just watch .. There are even those who live in the crucible of this blackness and live it with all its alcohol and depth ...!

I just imagined that my little brothers 'curiosity could make them open up to see it come true .. And I imagined my cousins' children at the time when their mother warned them that curiosity would make them spy on it and discover that it was on a site open to everyone, like YouTube , to listen and change for the worse.

In fact, this is a more severe epidemic than tuberculosis and plague.

Electronic Medicines-Digital Medicines-Introduction:
It is the first form of open control over the hearts and minds of the population through various means of the World Health Organization (WHO), the World Drug Control Committees and those responsible for the physical and psychological addiction to this type of addiction.

How many times have I talked about the problem of intellectual acquisition .. And this is a kind of its kind, but completely subversive and destructive! ...

These are audio files and tomorrow's television channels .. Instead of heroin, cocaine and instead of alcoholic drinks - vodka - bermut - bourbon - champagne - Italian, French and Russian wine.

Audio files interfere with the most accurate neural circuitry in the brain .. The effects on the brain are even stronger and more severe than those of a conventional drug that is sniffed, swallowed, eaten orally or injected through the blood.

The methods of drug use have varied so much from the early sixteenth century to the present day. but I think about how to punish her has become more common and more controversial among viewers, users or veterans of this kind of abuse. We studied in schools ways of addiction and caution, as well as ways to lure people into alcohol or use organic drugs to make users feel instant fake euphoria and happiness ...! There were drug addicts from all over the world, numbering in the millions, who were dying in her arms by the thousands . However, the World Health Organization (WHO) and the organization against drug addiction and drug abuse for many years to this day have been engaged in the arrest and activation of the search for their buyers and smugglers. Despite all this .. The wave of selling or promoting the drug is difficult to control .. Managers and owners of money are the main propagandists of drugs, and women are not only greedy for money .. Rather, it is a desire for self-satisfaction and the idea of absolute freedom from personality to society .. Therefore the question of arresting drug dealers became extremely complex .. Only the servants and retinue, who are primarily responsible for optimal lighting, fall prey to the head of the snake that controlled all these things.

So the story of drug addiction did not end, it got to the point where we found children who use drugs .. They are discovered by chance or through cultural awareness and take hundreds to juvenile homes for psychosocial and intellectual care and medical care.

Addictions are not limited to the use of organic and chemical drugs that slowly destroy the body, but there is more skill and greater ability than simple intravenous drugs .. It is easy for a child to become addicted to the network or older.

Digital Drug Trips
The journey to this digital drug is the name of one of the videos that, as soon as I started in the first second, I got so scared and I immediately realized that this is a devilish piece of music ...!

There are huge volumes in psychology that talk about healing through music and managing specific neural connections to bring a person to a sense of peace, harmony and calmness in cases of epilepsy, depression, or sometimes schizophrenia - minor schizophrenia - and, in some cases, schizoaffective hysteria. the scientific name of which is dissociative hysteria - or transformative hysteria - this type of advanced treatment is similar to the development of schizophrenia.Three- dimensional hypnotherapy .. This type of treatment is called stage III therapy for a person with esoteric hallucinations .. The patient sits in a half-stretched chair, and the attending physician speaks to with a very weak voice and certain words, and over time, the patient's hearing becomes very acute, but the brain moves from a state of consciousness to a state of unconsciousness .. Some therapists use other instruments besides a moving pendulum .. There is also a pendulum ring and other instruments used in hypnotherapy allowing the doctor to enter the vicious circle of an oblique idea of a patient who has lived for a long time under the influence, among certain nerves, gives him the real reality to which he was assigned.

Rhythmic musicology and others have a very strong influence on the course
of neurotransmitters coming out of the brain from the central nervous system in the brain
at the back and front of the head, which are thinly distributed throughout the rest of the
cells of the human body. You can either make a calm person live in peace, or be irritable ..
Many researchers believe that some of the noises trigger some sensory areas in the nerve
cell of the brain, forcing the brain to communicate more effectively between itself and the
person's mental state in the heart .. The heart makes sounds that alert the brain to
something .. Loud audible the music is either good or annoying, and this can be seen in the
gestures of the person himself .. He may develop a state of stress and anger, which
manifests itself in the movements of his hands or eyebrows and eye rotation .. And
gnawing teeth and lips .. Or he becomes very peaceful and retires to the kingdom of the
universe ...!

So I always thought that demon worship music is a very large part of a character's brain
colonization, in which the listener unconsciously harmonizes the letters and pantomime
used by the devil worship music .. And often .. The music is lonely .. And there are no
houses of poetry or matun written .. These are just auditory gestures that a person listens
to and receives the Curse of the Devil .. They are justified by the word "curiosity".

Just as music has a negative or positive effect on the movement of the nervous system of
the brain, which in turn is distributed throughout the cells of the body, scientists have used
ideal methods to get many people out of severe depression and after attempts to commit
suicide, there were attempts work hard and persistently.!

Many people with epilepsy and cluster tremor have recovered after listening to music .. In
addition, many cases of drug addiction have been treated at some stage with music
sessions, which give the person a feeling of some kind of euphoria, which takes him from
the stage of serious addiction to the stage of recovery from addiction. to become a normal
person .. He practices his normal life, full of love and optimism.

While we're on the subject of ultrasound therapy .. Only one American scientist has
discovered that this viable method should be applied in all psychiatric clinics to cases of
mild to moderate depression, schizophrenia or intermediate epilepsy.

There are also scholars who have subjected the Holy Qur'an to the same experiment with
a fresh voice and have given very amazing results ... Many cases are healed by reading the
Qur'an with certain sound layers that persist throughout the reading of the cases intended
for healing.

In this conversation, we are used to sounding "like atheistic scientists" - this is an integral
part of the cell .. It is natural that the empirical study of this issue is a new weapon with
which you can control the number of lives lost in this kind of electronic neurotic music.

How e-medicines work:
These drugs work in several ways, the most important of which is direct control of the
brain center .. And subject the entire nervous system to a more direct and faster metabolic

process than chemical drugs, which occurs through sound vibrations that produce nerve fluids that have no measurement, which the brain lets out to prostrate after listening to music, to feel an orgasm of some kind, similar to the orgasm of drugs, eaten or sniffed the same way .. But the next one is serious. :

A British scientist working for the World Health Organization (WHO) says something about how he returns from the dead and sees the unconscious in front of his eyes, being in another world. Let us mention a very important question about the awakening of a person who has entered a coma for a certain period and the way of caring for him, in order to return his internal organs to work without organic problems after the blood supply to the brain is cut off for a certain period after a stroke and enter into a coma for a certain period.

The doctor explains what I am going to write about now. :

When a patient who has been in a complete coma for a certain period of at least two weeks wakes up after suffering a severe traumatic brain injury, a "stroke", then, at least when a stroke is detected, some brain cells may be destroyed .. To avoid this, cardiac resuscitation is performed with electroshock so that blood is pumped further to the largest number of dormant brain cells!

When the patient is in a state between coma and initial awakening, we inject a stimulant into the brain cells to quickly and normally return to work .. But in most of these cases, the patient survives for three days, after which he has terrible physical setbacks, and then sudden death or death .. We decided to study the state of a brain cell before and after waking up, when it was free .. Any cell itself is a scientific experiment in which we prove the truth about what we did or not .. We discovered that after the cell is restored to life and then a stimulating substance is injected, which is the substance responsible for entering a group of neurons, that is, electricity .. It specializes only in the work of the nerve cell, which acts as a lightning rod on the heart muscle after how she stops living again .. At first we found that the cell is completely normal .. But after a while the cell the brain goes crazy, and I was very surprised when I saw her move with my own eyes so crazy and so fast ...! Boom .. The cell explodes and disintegrates completely .. And this scientific leap made us realize that if a cell is denied oxygen for a certain period of time and is given a surprisingly large amount of oxygen to revive it, it will not work well ..

This also applies to any part of a person .. Treatment is carried out in stages, so that the cells receive a substance that is injected to straighten part of it and return to its original state .. This is exactly what doctors in psychiatric clinics do, and then we decided to change a method of treatment for resuscitation of a patient with a stroke or stroke .. To be drugs that support the brain cell, gradually regaining consciousness .. Give in small amounts at regular intervals .. So that the brain cell can understand the variables and coordinates that have changed since the first period, when she stopped beating .. To return to life.

This is what the doctor said in one letter .. The stroke was saved .. He remained in a state of shock for at least ten minutes when the doctors knew about the seriousness of the condition and expected the patient to wake up with some kind of defect or paralysis in one part of the body or a problem in sensory and psychic limbs, but this experiment was a resounding success and resulted in it being completely destroyed.

From this story, we conclude that the mode and mode of action of this poison completely depends on the most accurate types of self-destruction of the brain cell, so that the cell is affected by sounds and vibrations of some kind that cause the brain cell to move in an unusual way and randomly release forced nerve fluids that suggest feelings of euphoria and happiness.

I expect this person to be very visible in the advanced stages, a week or two after listening to this type of audio file .. Since the addiction will not take more than two to three weeks, the following common signs will appear on the addict's face:

- Fading face

- Pale skin

- Features of stupor and lack of concentration

- Very slow assimilation

- Dark circles under the eyes due to the rupture of some blood vessels under enormous pressure on the nervous system

- Involuntary movements;. Tremors and others .. They are similar to the movements of some people with mental illnesses such as hysteria or neuroencephalic diseases, which cause them to involuntary movements of the hands, fingers or eyes, sudden movements of the legs, etc.

This is the last case of an electronic addict .. Which acts on a very rapid destruction of the brain nerve cell, which ultimately leads to its explosion and complete damage ..

The action of low atmospheric pressure outside an airplane at 30,000 feet above the ground is also similar to how an eardrum pierces when it hears some ultrasonic or infrasonic sound. the central and insane central nervous system itself is the same cell that leads to the explosion and its entire line and TV, which cannot be restored back even if I revived it. it does not react at all .. Because in the central nervous system one and the same cell was destroyed completely and forever .. The main carrier and receiver of sound vibrations was killed and completely destroyed.

And this kind of drug .. It is quite similar to some of the sounds used by professionals in the production and dubbing of horror films, specializing in demonic touch, magic and the supernatural .. While it is a kind of music .. It makes the cell feel a sudden sense of fear and then anesthetizes a sensory position in the cage so that the listener feels that what he hears is fun and fearless.!

This is the same method that specialists of the advertising system use in the issue of stereotypical influence on the brain of viewers of all advertising promotion.

If you continue to speak further, this topic will become a book to be studied scientifically in the search for medical or intellectual results and attributed to some scientists who have practiced or studied it before.

Frozen woman

Many people hope that their bodies will freeze so they can see the future and what it will be like, they are confident that the technology will become more advanced in the future so that they can be brought back to life again, just like with Jean Hilliard . who froze and then came back to life again, and her story is one of the best ..

On a snowy 1980 night in Lyngby , Minnesota, USA, Jen Hilliard (19 years old) was driving to her parents' house on a snowy road, and the temperature dropped to 22 degrees below zero, suddenly her car slipped and turned off the road, the engine broke down, and Jean tried to start it, but it was useless, she knew that at this time no one would pass by on this road, especially since it was covered with snow and ice, so she decided to leave the car, and if she stayed in her car and froze to death then this will be the best solution for her. is to walk.

Jin walked a long distance in very cold weather, and as she got closer to her destination, her body temperature dropped, her clothes were not suitable for such weather, and when she finally got to her boyfriend's house at 1 a.m. she was completely exhausted, her body was no longer could resist, and she fell ... The last thing she saw before losing consciousness was her boyfriend's house very close.

The poor girl lay in the snow for 6 hours at a temperature of 22 degrees below zero, and no one noticed her outside, who would leave their warm bed in such a frost! ...

The next morning, at seven o'clock, her boyfriend Willie Nelson came out and found her completely frozen on the doorstep of his house.

When Willie tried to take his girlfriend to the hospital, it was very difficult for him to put her in the car because she was frozen. He said, "I thought she was dead ... Her face was like a ghost."

Upon her arrival at the hospital, there was a big surprise for the doctors, where I had never seen such a case, they were surprised because she was completely frozen and cold, as if she had just been taken out of the freezer! Her face looked completely white, like that of a dead man, and her limbs could not move because she remained in this position for so long.
The doctors told Jean's parents that she had no choice but to pray that she survived. The nurses put hot compresses on her body to melt it, and her clothes were cut violently.

Dr. George Sater described her condition as follows: "I could not measure her temperature, because I had nowhere to put a thermometer, I could not open her mouth, I could not raise her hand, she was frozen and numb ..... Her joints were not talked, and even the eyelids, stiff and motionless, and the eyes did not react to light. Her skin was a hard needle penetration, later, when I could measure her pulse, it was 8 beats per minute. "

Two hours later, Jeanne began to suffer from severe spasms, this was a good sign for the doctors, but the danger still existed, they were worried about her condition after waking up, there was a high probability that the received part of her brain was forced to amputate her legs in order to save her life ...

"I took her hand and began calling her name, waiting for an answer ..." - said my mother. At one o'clock in the afternoon, Jean began to make some sounds ... He asked for water ...

By all accounts, it was a medical miracle, and that same night, Jane's arms began to melt, and on the third day, she was able to move her legs.

Jin said, "I woke up in the hospital and my vision was blurry and people were asking me who I was and stuff like that, and I couldn't understand why they were talking to me like that or why you were treating me like that. , you know these people and I know who I am, but I didn't know what the problem was? "

Sandra Jean's sister, who was recognized, said: "I knew that Jean would be, but an even greater miracle for me was that her legs began to heal and every time I lift the lid I find that the market has started to visit little by little , I still think it's incredible. "

Jean Hilliard left the hospital 49 days later, fully recovered, without losing any of her organs and living a perfectly normal life.
Jean Hilliard was a challenge for doctors to save her, and Dr. Ren Kelly described their success in saving her, saying, "It was enough for us that she was alive, it was good enough for her condition, but most importantly, not to lose one toe or one hand, only a few minor scars, which is great. "

But the question arises: how did Jane Hilliard survive ? ..

Science is always looking for logical explanations for accidents and strange things that happen, and there is only one scientific explanation. :

In an article titled "Is Hibernation Possible?" Published in 2008 in the annual report of child author Dr. Shang Xiao Li of the University of Texas in the Department of Biochemistry and Molecular Biology. "Some mammals can go into a state of extreme hypothermia during hibernation, as their metabolic activity is very low, and then regain full vitality when they wake up," he says.
In a study on the treatment of hypothermia, Dr. Lee discovered: "A natural biomolecule (5 AMP) allows for rapid metabolic failure in mammals. And maybe eventually the results of applied research and experiments in hypothermia laboratories will show us that there are huge life-saving opportunities such as injuries, heart attacks, strokes and many major surgeries. "

According to this study, what happened to Jane Hilliard was that she froze so quickly that her body overcame the tissue damage and went into hibernation, which allowed basic life functions to continue until they were successfully thawed out again. ...

And who knows .. Maybe someday in the future people will be frozen to revive them and bring them back to life at another time.

Relativity

This article may be difficult for many to understand, but I will try to simplify it as much as possible, as others do, posting some transactions that Valve considers not to study physics and math ...

Relative (1)
Comparative Einstein revolutionized physics, astronomy and the human universe ..

I won't talk about Einstein's life for a long time, but let's all move on to a topic that will hold you back, dear reader ..

Let's start together The Einstein of our study is inside:

1.Interpretation of the phenomenon of an electrode, which consists in the emission of electrons from the surfaces of objects, solid, liquid, gas, and when light hits them, electrons are also emitted by optical electrons.

2-contributed to the study of the movement-brown spots of molecules, as well as the random movement of molecules of the micron process (any particles are too small) in a liquid (i.e. liquid or gas). He came up with a method for treating the permeability (i.e. proliferation) of Brown molecules, where a relationship was established between the coefficient of permeability and the root-mean-square displacement of a granular Brownian. This way, I can establish the size of the atoms remaining in the fund, as well as the molecular weight in grams. He also can calculate the for and molecular number in his laboratory through an early relationship.

3-Time and Space: We will talk in detail about their nature.

4. We will talk about the dynamics of movement of objects later.

Now let's come together to understand, before we break this closed box, you must know what you didn't call, and to illustrate this, let's take a simple example ..

Relative (1)
During a car ride, we see trees around us, as if she is talking ..

Imagine, dear reader, that during sleep inflation the universe doubles everything around you, and you also double the size of you, my question now will cause this change? .- Of course not .. you will not notice anything, because nothing is absolute, it is still between you and the universe around you, each of you doubles with the same magnitude (force tour Henry one).

Another example, while driving a car and moving at a speed of 50 miles per second, we find the trees around us, as it were, moving backwards, despite the square, and if the car moves from our neighborhood, then with the speed B / C, too, in the same direction, as we find them motionless, but if we move in the opposite direction, then we find that it moves twice as fast as us, it all deteriorates, which means that nothing is absolute, and we do not see anything for real, and so on many, but is there anyone who can see the truth? Yes there is. let's take another example ..

Suppose a mass is tied to the end of a thread and placed on a turntable, and offered to two people, one of whom sits on a disc, which is an area of non-scarcity and stands outside, observing a disc called the reference area of palaces (regardless of the study of circular motion and generating a false one) ..

Why do you notice all of them, but is that so ?? ..

Will the person sitting on the disk notice that the cluster in the case of a square and spin, despite their rotation with the disk, while from the stand outside the disk, will notice that the book is moving and not static, about the wrong of the two? ..

The answer is nobody, and these two are eating right, but they all see books for him.

If the person who sees things, the absolute is a person outside the studied system. But the rejection of Einstein's idea of the range are reference, because there is nothing constant in the Universe, all the turns of the relationship with something else, am I now agreeable? ... You need to answer this question, it's something intuitive, I sit down to type, no static ..

No, this is wrong !! ..

Relative (1)
You think that you are sitting at a computer, but you do not know that you are moving at a speed of over 2 million kilometers per hour ..

In fact, I move with the speed of the Earth's rotation around me, with the speed of their rotation around the Sun, with the speed of rotation of the council in the Universe (600 km / sec) .. that was my question to any teacher-assistant in college: why not feel so fast? ...

She said that it was a long process, so if you want to know, I'll write about it in the comments so as not to go out on this topic.

If we now come to the conclusion that nothing is absolute, everything is forgotten by something else, the absence of reference to the palaces of reference, then what was

Einstein talking about in 1905, but if the place is relative, then is there time and also relative? ...

Yes ... time is also relative.

The idea came to his relative time to Einstein's, when you prove to Mickelson that the speed of light is constant in all directions, but when the plate is Einstein's idea of relative place, and the following example, it turned out that he should change the speed of light:

Suppose that: a fast moving car and someone standing is watching it from the outside, and the car has a lantern in front and behind, then when looking at a person it is necessary to find that the speed of light emanating from the lantern in front = the speed of the car + the speed of light.

The speed of light from the lantern back = speed of light - the speed of the car.

Relative (1)
Whenever we approached the speed of sound, time slowed down - the speed of light was 186282 miles per second. - ..

Any that the speed of light is variable. But Mickelson proved this with practical experience, proved the speed of light, so what's wrong? ... The unraveled puzzle that surprised Einstein so strongly, the reason for which lies in the time lag of objects to the body, when it moves, rapidly approaching the speed of light, where the slowdown of time until it reaches zero, and completely when the speed of the body approaches the speed of light. This explains some things about our bodies facing the light in paradise, no talking at the speed of light and time and staying in this world forever ..

Okay, so what if I were planning a body at the speed of light? . ((This, of course, will not happen.)) ..

The answer is that the will of the body overcomes the barrier of time and the past! ...

You may now wonder why we cannot reach the speed of light or the greatest of them, and in order to answer this question, you must know that mass is not constant, since objects change according to the speed of the opposite words of classical physics and that mass is constant. , as the speed increases, the mass increases, and when an object moves at the speed of light, its mass becomes infinite and stops against it, and therefore we rise to one of the speeds of light. But on every ruling street there are particles in the atom, called neutrinos, scientists thought that their mass is zero, but they knew that there is no mass, except for a non-specific (non-zero), this particle has a speed greater than the speed of light.

Einstein's view of time also crushed the concept of Newtonian reality, and he believed that time is absolutely and motionless, and Einstein will continue this way when he finds a connection between time and place through some simple equations.

When you look at the sky a year, watching the stars and the leadership does not know that these stars do not exist at the time you named it in it, it is known how the eye relies on the reflection of light falling on objects, therefore, to see the stars in the sky, there must be reflected light from the eyes, which takes a long time because the universe is huge and when this light changed its place.

The essence of the previous example is that space and time are relative to two, and it must be studied simultaneously and look at them separately.

Einstein's theory has been proven through various process experiences of many. It's like putting a watch in a jet plane, flying at a speed exceeding the speed of sound for several hours and then returning to the airport. It turned out that the watch, which was on the flight with a delay of a fraction of a second compared to the sentry, was on the ground and did not move.

In fact, there is another reason that can lead to a slowing down of time is the slowing down due to the gravity of the planet Pluto, which will help to stay on planet Earth for 248 years.

The concept (space-time) launched by Einstein is a term for expressing four-dimensional space (the dimensions of a place are three-longitude and height - in addition to the fourth dimension, there is also time). To determine the position of the body in a more specific way from Newton's zero space, a fourth dimension was added, which is the speed of light multiplied by time, for example, to add a fourth dimension to the equations of classical mechanics, to represent the space of racism in classical mechanics of relation, and then add the fourth dimension was calculated from the ratio.

The meaning of this word is they must change all the mathematical equations and physical properties of classical and quantum mechanics .. it all becomes wrong when applied to space, and some oddities of the relative and setting error of the law are recognized by the law of assembly, which is Newton's law of gravitation, which continued to operate for For 200 years, the law stipulates that when you place bodies at some distance from each other, each of them strongly affects the other.

$$F = Gm_1 * m_2 / r ^ 2$$

Relative (1)
Space-time warp .. science fiction image ..

And from Newton's point of view, this is caused by the rotation of the Earth, for example, around the Sun, as a result of the action of these forces in the center of the circular path of rotation, or rather it is an ellipse, as provided for in Kepler's three world laws, which discussed the trajectories of the planets around the Sun.

But Einstein did not give him reasoning that the reason for this rotation is the influence of the mass of the Sun on the curvature of space-time, it was believed that time and space are lead lines, put by any block in space on its curvature and corrected, which explains the *role of two small objects around large objects, but since this space can bend if the light does not travel in straight lines that determine the work of curvature, and this was proved by Einstein and found to be tilted at an angle of 1.7, and it was ...*

You may think, dear reader, that this is all nonsense, but this is how many phenomena of the Universe are explained, such as time, gravity and the shape of the gravitational lens. the last of my red gravitational delay times, my gravitational waves of gravity and oddities of orbiting Mercury ... and many others that could not be interpreted by Newton in any way (mind you, you can tell about all these phenomena in detail for another).

Before Einstein was born, there was a case of confusion among scientists about experiments with light and the theory of ether, which Dr.Mostafa Mahmud said about - this is bullshit and does not correspond to the reality of the disease, they also prove that through the famous experiment that made them spend years in search of a chicken about this theory, the actors who appeared Einstein, in which he denied the existence of the ether, which make his other scientists the absolute coordination of movement.

As the speed overlay in space has been proven, as explained by the example of the car I mentioned at the very beginning but refused to sign it, they said to keep track of the speed of light of one of the stars as it walked near the earth, and then smiled, but they found that the speed of light is constant in both directions.

*****

Relative (1)
Dr. Ali Mustafa Mosharafa ..

Before choosing a lender, I want to give someone the right, which is the physical world, the big Ali Mustafa Mosharaf, who is a theoretical physicist of Egypt, born in Damietta and Wales since the time of the throne. He graduated from the High Normal School and received his Ph.D. in Science from the University of London, and was the first Egyptian to receive this doctorate, he was appointed professor of mathematics at the High Normal School and then in applied mathematics at the Faculty of Natural Sciences in 1926. The title of professor at the University of Cairo is awarded before the age of thirty. Elected in 1936 as Dean of the Faculty of Natural Sciences, he became the first dean of Egypt. He received his nickname from King Farouk. He took as his apprentice a collection of the most famous scientists of Egypt, among whom was the book of Moses. After his exploration of the world, Einstein described his theory of relativity and described him as one of the greatest scientists of physics, and his research on quantum theory and the mechanics of radiation from maize about 15 searches and drew up a scientific research project just before his death about 200 projects. Dart his research project on the application of tape, and the quantity is normal, which will allow us to explain the phenomenon of he Zishan.

This was the doctoral first project of scientific research on finding a measure for the brain, where the geometry of vacuum, based on the theory of "Einstein", concerned only the motion of a particle moving in a gravitational field.

He studied the project of the relationship of money and formulated a scientific theory that is important in this area.

The doctor was a project of believers in the importance of the role of science in the progress of peoples, and his victory among all sects of people, even if they did not get rid of it, so his attention was focused on the development of generalizing principles of this science, on the knowledge of a simple layman who could understand them and which Like any other thread, they were constantly mentioned in the introductions to his books, explaining the mysteries of scientific knowledge simply and clearly understood by all people, even non-specialists.

Most important of his books:

- Mechanics and scientific research 1937 - Descriptive geometry 1937 - Slang of airport communities 1943 - Plane geometry and spatial 1944 - Trig planar 1944 - Atomic and atomic bombs 1945 - Science and life 1946 - Geometry and trigonometry 1947 - We have and 1945 - Special theory of relativity 1943

Relative (1)
Most likely, he was killed by the Mossad at the age of 51, and later raped his student with the stuck corn poison of Moses, while in America the 34-year-old accident was staged in a big way.

I was raped by the project when what Einstein said: (I can't believe the project is dead, it is still alive thanks to its research). He also said on another occasion: (he was the project of the encyclopedia of a man, a whale of great talent, who tells the same all over the world). And he said in another hadith: (the mind was a draft containing many of the secrets of the atom, but fate wanted him not to reveal them). He also said: (on the day he died, half of the flag) on the day of his death.

***

I think I had quite a lot, so I will dwell on this point and promise you that the second topic discusses some of the uniques arising from Einstein's assumptions and the application of these hypotheses in our social life, as well as the phenomena of the Universe, which were explained by Einstein in his theory, as well as we will consider the division regarding the quality of it (private - public) very briefly, since I have to regretfully outline some of the transactions that Einstein proved to slow down time.
You have to stop to think, and let's talk carefully about the illusion of time.

In this article, we will try together to define the meaning of time travel? .- Could you travel in time? If this is possible, how? ... And some unusual results from this theory.

But how can we talk about time travel if we don't know what time it is yet? !

Commentary: What a time! We all know that time is just the ticking of a clock, every second is past, present and future, and could a constant flare up that does not change when I look? She is one second after the middle of the night ... but where ??? ...

I think we should stop to think, and let's talk carefully about the illusion of time.

I mentioned in a previous article that Einstein contacted a relative of his time when he discovered a subtle connection linking him to the office, and that times change with your speed and where you are from Earth.

Change the time to speed (we talked about this in the previous section).

Times change as a result of office work (demonstrated) using modern precision nanoseconds, where it was found that the clock slows down whenever I place in the gravitational field the strongest change in the time subject's gravitational field is due to the deformation of the Earth's case by the tide.

If time is not fixed, it may speed up or slow down, but it spreads very little, I don't feel it.

I said in a previous article that Einstein added the dimensions of the spatial fourth dimension - this is time, and saw that the decade is a place and every moment is twice in the past, and every moment we live now and in the future as needed.

But can you really move from a point in time in the present to the future or the past?

Before you mention the various means that scientists have tried, made a time machine and travel in time, which I see stranger than fiction, you must remember that time travelers exist among us, who are they? ..

Quite simply, they are passengers in space outside of Earth's gravity, scientists and others simply move them out of the Earth's gravitational field rapidly as time passes over Earth's time.

Examples of numbers
Relative (2): time travel .. strange results
Someone flew into space, and a second shadow fell to the ground.
The twins at the age of 10 left one of them on earth and went on another journey into space, and suppose that the gravity of the wheel came in the traveling wheel and some accurate calculations we find that six months on Earth comes 6 weeks for a child-passenger in his rocket at a speed that, if the rocket travel continues, the duration of two becomes the age of the traveler 12 years, while the age of the brother on Earth is 38 and varies with the speed of the rocket .. (comment: time travel, draw and mesh travel, what kind of students Oh, these sick schools: D).

Only a 28-year-old child can travel to the future through a time machine and only a rocket.

This is the equation that did the calculation: $t = t_0 / \sqrt{[1 - (v^2 / c^2)]}$

Now let's look at some ideas, Very strange, written by many scientists in the form of detective transactions in theory, but ended up mostly in failure or inability to investigate in practice:

one . the idea of building bridges around the universe

The idea of the mathematician Kurt Gödel, who came to the world, was the idea of curving space from space-time (the previous section) and the transactions he formulated and mapping them to Einstein, who saw it as theoretically correct, but that was the imposition of the rotation of the universe around its axis, which is wrong in theory.

2 - the idea of building a rotating cylinder

The idea came from one of the physicists as a result of the lack of rotation of the Universe, let's show them that we are processing the Rotor, as a result of which we make a loop in space-time that allows us to travel (movement in the form of a spiral) and of course we formulate equations in theory, but , unfortunately, he needs a finite cylinder and that, of course, the future is moving.

3. The idea of a black hole (worm hole)

Relative (2): time travel .. strange results
The bodies of low-density black holes are very high.
You can't build a rotating cylinder, but can't you even build one? ... There are objects that rotate very fast, so they are black holes ..

What happens if you enter a black hole? .... Quite simply you will die, "baby": D

But what are these black holes?

It is very simple that the density of small objects very strongly distorts space-time because of its attractiveness, and now the idea of travel-through, a finite space-time has come to create a closed path between two points separated in time and space, but according to one of the scientists, that The black hole is only a one way street when you never leave it.

But some scientists who don't like this failure have come up with the idea of a new wormhole, what is it?

It is a short connecting point of two different times and far in between, there is no evidence of its existence, but the general period sees that they are small in size and maize, and it should be expanded into motion and a signature made by anti-gravity energy called negative energy .. (comment: wormholes are weirder than fiction, but it should be mentioned).

4-idea of cosmic strings

In the realm of thought, making a journey through it with the help of the existing energy on cosmic strings, the thought came about the impossibility of making a spaceship move at the speed of light and at the same time receive short cuts through it at a speed slightly exceeding the speed of light, they say the same thing that and the distance that is light, but in short then we kiss.

An important question - What are cosmic strings?

Scientists speculate that they are threads of very subtle energy that occupy the universe, but in reality this is just an assumption, and no one knows if it exists or not.

After all these roads of strange and applicable work, I can say that travel to the past is still impossible.

Time Travel in Quantum Physics: I will not go into this topic (it is difficult to explain through the topic).

Time travel from the point of view of philosophers: I won't go into it; I don't really care to know if you traveled back in time, how you killed my father, is old or not speech, which is useless to the nearest person from the floor and gives answers ..

I will stop here with time travel and converting the apartment to a different location and this results in an exotic look.

I'll start with a very famous example:

1 - example train / Note: (all following numbers are assumed and process)

Relative (2): time travel .. strange results
Suppose the train is moving at great speed ..
Suppose a train moves at a speed of 240 thousand km / s and a length of 5 million 400 thousand km, and suppose the existence of a lantern in the middle of the train Babbit-cats in front and behind open automatically as soon as the light hits them .. And when you stand and love that , what do you see? ...

In the first case, and you are sitting in the middle of the train, you will see a light moving at the same speed in front and behind the train, and will automatically open the doors at the same time after 9 seconds, because the train for you you are inside it. its speed is zero.

But what do you see and you stand on love?

You will see the light moving as if it is trying to catch its arrival train at a high speed and threw it after a 45 second break, because it sees the light turning to superimpose two speeds because they are in the same direction 300 thousand (speed of light) -240 a (speed of a train). While he sees the light moving in the direction of the oncoming traffic at the rear of the train and thus to the door at the rear in 5 seconds, only because he sees that the light turns are gathering two-speed 300k + 240k.

But hey, what does it all mean? ... Like a door that opens to a person inside the train at the same time, and someone else outside the train at different times, I think we are on the verge of insanity ..

2-relative lengths

Believe it or not, the fact that the length is not fixed has changed too. (Note: proven in practice, not just imposed)

*The change in the speed of any object is accompanied by a decrease in length, the faster, the more it swings even before the complete disappearance of the speed of light, depending on the ratio L = lo * sqrt [1- (v ^ 2 / c ^ 2)]*

3.gravitational lens (one of those phenomena that Newton himself cannot interpret by gravity)

Relative (2): time travel ... strange results
Albert Einstein is the greatest scientist of our time.
I often wondered why we see stars in the sky that are too far away, and our star next to which is supposed to obscure vision by its large size (not because of the light of which), the answer came .. this is due to the presence of celestial bodies , larger than the Sun hundreds of times, which in turn bend the light as a result of their appeal to very large, which in turn are reflected on the celestial bodies of others and thus appear on Earth, which is not so many pictures through a telescope.

I will focus on the weirdness of relativity and talk a little about Einstein's own weirdness.

- Einstein studied in college at the age of four, but he did not answer him until he was nine (at least so his parents thought), until one day at lunch he said that this soup was very hot, and when he asked his father why he did not speak all this time, he replied: "Everything went on until now."

- Dismissed from school for poor performance in the university entrance tests, as well as for poor performance in all tests, except physics and mathematics.

"He hated shaving and changing socks.

- He proposed to the woman and her daughter at the same time.

- He would like to become a violinist. if not physically.

"After the death of his stolen brain, studies were carried out to find that the cell formations were different from those of an ordinary person.

Dogon Tribe Secrets

The Dogon, a large African tribe in Mali, numbering more than three hundred thousand people, live on a rocky plateau, live a normal and peaceful life and organize seasonal festivals characterized by an amazing variety of colors, dances and rituals.

Dogon secrets

We can consider the inhabitants of this tribe as ordinary people before we hear about their amazing cosmic knowledge, where the high priests there possess science and astronomical knowledge, which raises questions and bewilderment, despite their primitive life, but their religious culture is completely dependent on Space Science and specifically from the star system (Sirius), which can never be observed with the naked eye, where they have a satellite.

Dogon legends

The life of the inhabitants of the Dogon is full of legends dating back to time immemorial, where these legends speak of the origin of this universe by God (AMA), who created Earth granules and launched them into space, and then turned them into stars, and then two white clay granules followed and the meaning here (Sun and moon) and ended up with a granulator made from a column of clay (earth) and from the same granulator.

These myths may seem common at first glance, but they become interesting when you talk about these eight and say that they originated from a star called (Sirius), and that this star is accompanied by another star called (Sirius B), where absolutely it is unclear what they know about this star, which was only discovered in 1836 and named after a white dwarf in 1915.

Dogon Science

In 1931, two French anthropologists (Marseille and Germain) went to live with the Dogon, then called French Sudan, to study their culture, having lived with them for 20 years. At first they were not welcomed, but over the years they were able to win the respect and recognition of the inhabitants of the region, and in 1946 the elders of the tribe agreed to teach them cosmology.

The priests held sticks and drew on the Earth the Shape of the sky as their culture intended, and then both worlds were stunned when the priests began to draw a large object around which a smaller star revolves, scientifically known for the fact that this star revolves around the star Sirius once at fifty, and it is surprising that every fifty years the Dogon celebrate a holiday called In their language (Sirius B) he calls b (photo), i.e.

When two scientists asked about the source of this information held by the priests, they told them that it was reached by amphibians, mermaid-like aliens named Nomo, who landed on Dogon land a long time ago, and they claimed that these creatures were the Guardians of the universe. and the fathers of humanity.

Almost everywhere, the Dogon draw figures that describe the arrival of these creatures on earth and pinpoint exactly where their ship landed, indicating that it has landed in the northeast of the Dogon, near the rocky plateau where the inhabitants settle.

Like all legends in the world, there are many more of their symbols that are unclear or understandable, but one thing is for sure: the Dogon's knowledge of astronomy far outweighs their knowledge of observation or calculation.

Dogon and astronomy

After a while, having learned these secrets, the scientist Marseille discovered that this tribe has many other modern astronomical sciences that they have known for many centuries, for example, they know that Jupiter has four main moons, that Saturn has rings around him that the Earth revolves around the Sun and that the stars are constantly moving bodies.

They also know that the Moon is a dead planet and they know about the war of the spiral Milky Way that belongs to our solar system.
Another fact they know about Sirius is that it has a second star that is different from Sirius B, but until now it has not been discovered by astronomers, and if it is ever discovered, it will confirm the Dogon myths about their contact with aliens.

The question that arises strongly now is where the Dogon got their knowledge from, that is, what is their true source, far from their claims ?! ...
Assumptions about the Dogon mystery
The speculation of an alien landing and exchange of information with them certainly sounds unconvincing, despite their bizarre description of the ship, where they claim it caused a sensation that shook rocks off the ground and sent dust up into the sky.

Unfortunately, these stories have been passed down orally from generation to generation, so the likelihood of losing or changing information will be very high and similar to other myths filled with vague or persuasive symbols that cannot be studied or given a logical mental explanation. Still, there are those who supported the idea of aliens, such as the scientist Robert Temple, who wrote his book on the topic with great daring in his conclusions.

On the other hand, there are those who reject this interpretation and give another, simpler possibility that the Dogon were trained by a French secular school in 1907, so their science is entirely derived from French civilization.

While this argument is very logical, it may unfortunately be wrong because the Dogon myths existed before the French colonized them, and furthermore, it is unlikely that teachers at that time were giving the Dogon astronomy lessons, especially in such precise questions. like a star (Sirius B).

Another, more likely hypothesis, is that people in ancient times were more inclined and loved astronomy and Middle Eastern religions full of such items, the Dogon tribe was not isolated from the world, but had trade routes even to Egypt, so it is likely that they drew their knowledge from ancient religions, whether in Egypt, Mesopotamia or even Greece. It is also known that the star Sirius is the brightest in our sky, the designers talked about it and associated it with the flood of the Nile for the first time, and on the other hand, many Greek myths speak of amphibious creatures, half-humans and half-fish.

But another question arises: the star Sirius B cannot be seen with the naked eye. ...

Some speculate that in ancient times this white dwarf shone in the sky and people could study it despite their primitive instruments, and in fact many ancient civilizations were well versed in astronomy.

Output

This tribe remains mysterious, and there is no final and final word about the origin of their legends, but if one day the star of Sirius III is discovered, this will inevitably force us to reconsider our hypotheses.

Guests behind the stars

In 1973, the Governor of Georgia invited 20 of his guests to dinner, and when they turned around the table, they saw from a large glass window what caused confusion and fear in their hearts.

The body of a large bird, talking, changing its colors.

The Governor said: If I become president, I will announce the secrets of UFOs, because I saw one of them!
This man is Jimmy Carter, who later became the 39th President of the United States, and everyone expected him to have his say, but he said nothing.

The case against the anonymous author has been ongoing for all these millennia.

If we try to examine the file from the first page, we will be confused, what is the first page ?!

"The priests told me that the sky of the country was filled with shiny silver objects, and it was in the winter of the 22nd year of my reign .. I ordered this phenomenon to be recorded for posterity!"

It is a papyrus text preserved in the Vatican Library, dating from the era of the pharaohs of Egypt, on which supporters of the UFO theory are based, moreover, it has existed for thousands of years.

The Old Testament itself is a ceremony with similar references. Another document, dating back to 1180 in Japan, is about a "flying saucer" seen over Mount Kiei.

Even the roots of the term "UFO", swirling around rippling differences, were the beginning of takaka in 1948 on the lips of a pilot (Mantel) who shot down his plane with a body unable to describe, so give it that designation ?!

Pages and dates continue in a row.

In 1235, General Yeritsumi couldn't figure out if what he saw was true or fiction!

Glittering balloons flew over Japanese Kyoto, and if his army hadn't confirmed the same, he might not have believed his eyes.
In 1561, history recorded collisions between UFOs in the skies of Germany in what was known as the Nuremberg Incident, one of the most common incidents collected by a large number of eyewitnesses.

Papers have accumulated in the UFO File, to double its size more and more, we select from it a page of a certain degree of reliability in the circle of British knowledge, and to be precise, this is page 855 in volume 18, the page contains the so-called "Brazilian incident" , in which a UFO exploded .. Its parts were made of ultrapure magnesium + strontium!

This is an alloy that is difficult to manufacture using our technology.

This time, the incident is contained in Timothy Good's Top Secret, pp. 368 in a book, this time aimed at the New Mexico incident, and How Rafting Remains does not exist on American or Soviet rafting maps.

The number of events and views doubled more and more, and in 1952 the United States could no longer occupy a spectator position, so they launched the spark of (Blue Book) project, in which they flew to analyze thousands of these news, and it turned out that only 6% of them are the result of an unknown phenomenon, and the rest of the collective illusions have a thousand reasons and reasons different from the usual .:

- Meteorites.

- Thunderstorms.

- The gases rising from the swamps, the element methane contained in them, is known to be flammable in nature, and it is easy to achieve this state with thermal impregnation.

"Complex aircraft are being tested, and the US military is said to rejoice at rumors of UFOs because of this, distracting from any viewing of their modern flight experiments.

The Army even admitted that the Roswell incident in 1947 was based on an observation balloon as part of a secret project to control radiation levels in the event of a nuclear war.

Let's turn to the 5%, which is very hard to deny, and most of it is backed up by collective testimony from trustworthy individuals, as well as confirmation from electronic sources such as monitoring stations or radars.

The most famous example was in 1986, when passengers on a JAL plane could clearly see a flying saucer while a radar station simultaneously observed it.
Ten years ago over Tehran and observed by thousands of Iranians at 11 o'clock in the afternoon, Sir Arthur Clarke recalls the incident in detail in his book "A World Full of Mysterious Secrets", and how the Minister of Defense was among the witnesses and sent two aircraft to intercept it, one MiG 21, and then another MiG-23, two other MiG-23s.

One of the lawyers filed a lawsuit against the U.S. National Security Agency with the number: 5U.S CODE-Section552b1, which required people to disclose their secrets about UFOs, and was based on federal law in the states that allows a citizen to access any information until then. as long as it does not affect the security of the country.

This law is called FOI A, these are my initials: FREEDOM OF INFORMATION LAW

The head of the agency said that he could tell the judge only a fraction of what he had, and after a closed bilateral hearing, the court's decision was made:

"The documents in their entirety ABOVE Top Secret Page 413, if made public, seriously harm the interests of the United States, have been postponed."

Time Travelers -2

Physicists such as Albert Einstein and Stephen Hawking say that time travel is possible and feasible, but the problem is that science has not been able to achieve this. But can Nature do it?

Towards the last

Mr Elsie and his business partner Charlie finished lunch at a restaurant in the small town of Abbeville in southwestern Louisiana and then left at 1:30 p.m. heading north on Highway 167 to Lafayette, about 15 miles from here, where the oil hub is located. continuing to conduct business negotiations.

It was October 20, 1969. The sky was clear and blue, pleasantly cold, and of course both men had all the necessary facilities to roll down the car windows as they drove towards their destination. The car drove along the highway without any obstacles, completely motionless, until they noticed an old Volkswagen at a fairly close distance, moving slowly in front of them. And when you kill more than this car, they turned the conversation from the work area to that old car, which noted that, although very old, it was in good condition, in a manner that made all men admire.

They decided to overtake the car because it was slow, but as it got closer, Charlie saw that the speed had decreased to reflect the car and its excellent condition, and saw a bright gold plaque clearly printed on it in 1940. This was surprising, in addition to what was considered illegal, if the owner of the vehicle did not issue a license giving him access to the old number plate for ceremonial parades and national parades.

Charlie passed the car to the right of their car, and Mr. Elsie, in the passenger seat, noticed a young woman at the wheel of this old car, dressed in what looked like an old 1940s dress, with a fur coat and a long flowered a hat with a feather is a dress that people in 1969 were not used to seeing on women.

On the seat next to her was a little boy or maybe a little girl, the gender of the child was difficult to determine because he was wearing a very heavy coat and hat. The windows of her car were closed, which puzzled Elsie, although the temperature was cold and cold, it was pleasant, and a light treasure would suffice.

The attention of both men turned to the look of fear and panic on her face, and the lack of any movement in both directions allowed Charlie to slowly approach to stand next to her

so that they could see her looking terrified, as if she was lost or urgently needs help, and she seemed about to cry, and tears were about to gush.
Elsie, who was closer to her and shouted loudly, asked if she needed help, and she nodded in the affirmative, looking at their car as if she had seen her for the first time. He told her to stop at the side of the road, and he had to repeat this several times as he gestured with his hands and tried to utter words on his lips because the windows of her car were closed and it was difficult for her to hear him. Then they saw her parked at the side of the road and overtook her to park quietly in front of her.

We got to the beginning of the road and stopped our car, and when they turned back, it was no longer an old car, there were blows that hit all the men, because they saw a woman standing with her car in front of them, and then they all ended up on the highway, where there are no side roads, and nowhere along the way can you hide the car. The car simply disappeared along with its passengers.

Charlie and Elsie returned to their car, perplexedly looking at the empty road, and it became clear to them that it was impossible to find this car. Meanwhile, the man who was driving behind the woman stopped, ran up to them and asked in surprise what had happened to the car that was driving in front of him. He told them that he was driving north on Highway 167 when, at some distance, he saw a new car slowly overtaking an old car, then he saw the new car stop at the top of the road and the old car stopped, and for a moment the vision was obscured by a new one. machine, and then suddenly disappeared.

The man was desperate to find a logical explanation for what he saw, so he assumed that the woman had an accident. After the three men discussed what they each saw, the men combed the area for an hour, then a third man, who was from out of state, insisted that the police be notified of the incident, saying that he felt that this missing person case they witnessed, Charlie and Elsie refused because they had no idea where the woman and the child were.

In the end, the man decided that without their help, he would not be the only one to report what he saw to the police, fearing that his mental health would be questioned. And he exchanged phone numbers, addresses with Charlie and Elsie, and stayed in touch with them for a long time, and they talk through their contacts from that incident, and he assures them that he is sure of what he saw.

What if?
What if this woman was from the past, came to the present, and is now an old lady? But what if on that day she was driving in her car instead of Charlie and Elsie behind an old car, was this lady the same ?. What if he came from the past to the future and did not return to her past? Would the newspapers of the time confuse the woman who disappeared with her child on a cold October day? And the search for her continues as she and her child travel in time and back forever.

Witness an air raid from the future
In 1932, journalist Bernard Houghton and photographer Joachim Brandet were tasked with conducting a press investigation into shipbuilding and refurbishment in Hamburg,

Germany. They drove there, talked to some of the leaders and workers, and then finished their assignment in the late afternoon.

As they were about to leave, they heard the unmistakable sounds of drones as they looked up to find the city sky covered with warplanes. They then heard the city's anti-aircraft artillery fire as bombs went off around them.

Moments later, the area turned into a raging hell when fuel tanks were bombed. Dock area warehouses collapsed due to exposure to explosive materials, and dock cranes were turned into pieces of scrap metal. Hatton and Brandet realized that this was not military training. They rushed to the car while artillery bombed enemy planes, which sent bombs from the sky. When they left the gate, Hatton asked the guard if they could help with something, but he asked them to leave the place immediately.

Brandt took photographs throughout the airstrike, and after he showed them, there was nothing unusual about them. The images showed the shipyards as they were when they arrived there in the morning, there was no sign of a rain of explosive bombs dropped by any enemy aircraft to destroy the area as they saw.
The newspaper editor studied the photographs taken by Brandt, but was impressed with Hutton's insistence that they witnessed the air raid and photographed her, refused to accept their story and suspected that they stopped at a bar and drank before returning to the office. which made them imagine a plaque.

Hutton left for London shortly before the outbreak of World War II and settled there. One day in 1943, he read the newspapers, saw in one of them the news of the successful raid of a squadron of aircraft of the Royal British Air Force into the Hamburg shipyard, and he shivered when he saw the photographs. The scenes of destruction were exactly the same as those he and Brandt saw during their visit to the shipyard in the spring of 1932.

There was only one difference: Hatton and Brandt had witnessed the incident eleven years before it happened.
The White house.
A Ford pickup truck carrying Carl, Mike and Gordon arrived at a cattle pasture near Buncan, Oklahoma in the early fall of 1971 and pulled up at the gate. The three men worked for a livestock feed distribution company that sent them to this remote area to pick up the feed container from there. But what they saw there silenced them for 41 years, the gate was just barbed wire with no locks, they opened it and then entered a private property that was covered in weeds growing out of negligence and obsolescence, so that it came and bypassed the hood of the truck, and yet the men drove their truck through this long grass until they reached a tank parked next to a, but they could not carry it to the truck because it was half full, so they decided to leave and let your boss know.

As they drove out in their truck, they rounded the shed and saw on the hill in front of them a large white two-story house with no lights on.

All three returned to the feed company and told their boss that they could not carry the tank, and later in the evening the boss told them that he had emptied the tank and that they could return for it tomorrow. They did go back there the next night, and Karl says, "We decided to take weapons and check the Old White House when we get there."

They drove through a private property on the same road they drove yesterday and then analyzed the tank, after I was done with that, they drove their truck around the barn. But what they saw was engraved in their memory for many years.

"We climbed the hill, but it was not the house that we saw there last night, there is no sign of its demolition, there is no sign of its rules, there is nothing at all, what we all saw last night no longer exists, we talked to each other over the years, but none of us could provide an explanation for what we saw).
Four years before the future
Young Jake stood outside his parents' house on the Lake of the Ozarks in Missouri at 10 p.m., leaning against his truck, staring at the northern lights coloring the sky on May 28, 2004, unaware that his life was about to change forever.

Shining with a white light in the sky that filled the entire northern horizon, it never looked like the red and green lights of the northern lights that Jake read, and it didn't shape the way you do. This light moved like the light of a photocopier as a streak of bright light passed over Jake's head, moving from West to East, and then disappeared. When Jake saw this, he thought he needed to enter the house to be safe, but when they realized he couldn't do it, his arms and legs were numb and then he passed out.

He was unconscious for about an hour, and when he woke up, he felt disoriented and almost numb, unable to understand what time it was. Jake returned home and it took him most of the night to tell his parents what had happened. He was convinced that there was a mistake in the calendar, and all the time he persuaded them, insisting that the year should be after 2008.

"To this day, my mother remembers some parts of this incident, mainly because, in her opinion, one point of view is that I asked the question:" Is the president a black person? ".
Strange guy.
Kell walked out of the store at a gas station and why are they opening the door of his 1999 Chevrolet, which he parked next to the store, and when a very large man with a head the size of a watermelon suddenly speaks to him, Kell exclaimed, "What year is it?" The man was standing where Kell had walked past a few seconds before leaving the store, but he did not notice that he was there, he was wearing a black business suit made of raw fiber fabric, and Kell says it was a type of fabric worn by Teddy Roosevelt. "The man was in his 30s or 40s, tall, white, clean-shaven, he looked normal but asked a strange question," Keel said in 2003, but his face contorted with anger and he shouted sharply, "What year is this? answered (again, this is 2003), but he repeated the question again, and Keel replied: Kell looked at the man when he got into the car, and when he sat down, he looked at him again, but he disappeared.

The man disappeared from the front of the gas station within two seconds it took Kell to get into his car, Kell looked into the store, which was the only place the person could go in those few seconds, but it wasn't there. He just disappeared.

Theodore Roosevelt was the 26th President of the United States, and his reign lasted from 1901 to 1909.

Telekinesis .. From scam tricks to scientists' laboratories

In the wonderful Stephen King (Carrie), a shy schoolgirl has latent telekinetic powers, and on graduation day, her classmates doubled down on the pranks they throw at her, forcing those abilities to come out, in a clear message to us readers to warn our introverted classmates a thousand times over. at school.

It is not surprising that this novel made the king famous, turning from a writer immersed in the king of horror literature, as he is called, as for the movers in the world of reality, then perhaps the most fortunate in fame is the Israeli Yuri Geller, who has long headed television screens. and he gladly praised the commentators only with his spiritual strength.

A large number of skeptics remained a visual gimmick until Geller appeared on the Johnny Carson Show in 1973, and the host suddenly replaced the commentators on air that Geller had equipped with another group, and this is where The Legend broke for the first time and Geller was unable to bend it.

The problem is scientists can lose to Hawa, they are fooled by their sleight of hand tricks, and there is no atom of superhuman strength.

Even the term "Psychokinesis" was invented by scientist Michael Talbourne in 1982 specifically for two boys who struck him with their telekinesis, Michael Edward and Steve Shaw.
A year later, both boys admitted that they were resorting to tricks, not possessing real hidden abilities.

While the term has struggled to survive, Psychokinesis, whose origins are mostly in Greek, means movement under the influence of the brain, while many academics abbreviate it to PC.

A number of experiments were carried out under close observation, and one of them, after a strong effort, barely had time to move the thermometer by a tenth of a degree.

True, this is a small success, but it is real and proves that this is possible, since there is a famous video of Nina Kolgina, in which she separates the yolk from the egg without touching the glass vessel in which it is located, at those moments the ECG recorded that her PULSE reached four times the normal rate, about 240 beats, the video was filmed.

A controversial video about Nina Kolagina's abilities ... Watch the video carefully, watch its capabilities and judge for yourself ..
Scientist Michio Kaku believes that the future will be with machine-assisted telekinesis, in the sense that, naturally, a person can control the brain by focusing on a certain idea, you can send these ideas into a computer, which translates them into motion.
Neuroscientist John Donahue made a breakthrough in this area by inventing the [Brain Gate] "gate of the mind" that helps paralyzed people move from the same scientific perspective, and his results exploded when one of his patients appeared in publisher magazine in the summer of 2006 and he was able to open his email with his mind, change TV channels, and also play videos.

Definitely going to make these successes a window of hope for many, but in my opinion I do not think that this applies to controlling your remote in the sense that he knew it, but it is a process that begins by skipping the highlighted with the machine, and at the stage when the second machine is in motion.

In the near future, telepathy will not become a supernatural thing limited to some transgender individuals - as shown in the X-Men movie - Perhaps everyone will be able to read and share thoughts with others without problems ..

Scientists from the Institute of Molecular Manufacturing in California have developed the concept of implanting human artificial heads - clusters of nanobots, micro-microscopes that can be inserted into the human brain ..

These small particles transmit ultrasound data to centers of activity located in the nerves of the inner ear cavities, causing the recipient to feel a sound from inside their head that only he can hear.

It can also be developed to the point of showing (video) in the retinal field of view what another person sees.

The scientists involved in this project say:

If the necessary funding is provided by the appropriate budget, then the so-called telepathy will become available to everyone in a maximum of forty years!

Between science and Maori .. What do we believe in?

Croup (sleep paralysis)
We all know this strange phenomenon, which people sometimes call "sleepy madness", but if you don't know it, let me introduce you to it.

Croup: This is a feeling of heavy weight on your chest that prevents you from breathing, moving or speaking and thus causes paralysis and the feeling of someone sitting next to you or on your chest.
The corpse is old.
Religious and paranormal scholars with an interest in the pursuit of ghosts and goblins say the phenomenon is one of the mysterious cases of humans, especially since its early appearance in the Middle Ages, when some people developed cases of anxiety during sleep, paralysis in motion and shortness of breath and has been classified by physicians as having been caused by eating earlier. Well, this is for the rich and the noble .. But what about the poor? .- Do they eat the way rich people eat? ... The priests said that this is a consequence of complete disbelief, lack of piety and adherence to religion .. And therefore God does not stone people .. Some of the priests themselves were infected with the Jatom disease, and at that time people doubted the reliability of their statements if they were honest ...
Now sit down.
Since science is the only square in our century, it had a firm and convincing answer to the words and opinions of those who inherited them from the ignorant generation. With our entry into the TWENTY First Century, everything that scares people has become just a joke and is ridiculed by some, because scientists have responded to them with scientific evidence from sources and research .. After analyzing the phenomenon of Jatom, it turned out that it was not caused by jinn, demons or disbelief and not paranormal phenomena of

any kind of interference on the meanness of the human body .. The thing is that the brain relaxes the muscles to achieve maximum sleepiness, and this relaxation gives rise to general paralysis and inability to move.

Scientific studies of the brain have proven that sleep paralysis is a normal state, when the body is asleep and bouncing, naturally, you see that you can neither move, nor breathe, nor speak, because your mind in this case is a razor on the horizon, and you are in your bed.

The reasons for its occurrence:
- Curvature of the septum - sleep apnea - sleeping on the stomach or face - irregular sleep schedule - lack of oxygen in the brain - human use of sleeping pills

Solutions
Doctors advise you to exercise every day rather than go to the gym - all you have to do is move your joints and muscles to activate circulation.

Spirit preparation
One of the most famous myths of the ancient era, it is considered one of the ways of communicating with spirits, invoking the spirit of the dead and communicating with them through mental or spiritual telepathy .. This phenomenon spread around 1700 with the advent and spread of tales about the werewolf, witches, sorcerers and vampires, and their history may be older than the history of these sessions, but this phenomenon hangs in the air between rejection and acceptance between those who like to deceive themselves and those who have a sincere opinion.

Preparatory studies in the past.
Usually one of the Mages sits upright, and then a poor client, who does not know anything, comes and asks him to call the soul of one of his relatives so that he can be reassured or asked about something. The magician then takes some details, in which he uses his sharp mind after a while to convince the client of the answers he wants to hear. The magician comes up with events and details, evoking the soul, and imagines physical awe and serenity in his mind, and then says, "He is your relative here .. Ask him and I will hear him and answer you with what he has to say." ... And so the scenario went, and from the heat of ignorance people crumbled on the threshold of witches and necromancers, nobles and princes, and novice witches, and even the poorest of the poor .. This prompted the phenomenon to stand out on the surface of history and events.

Preparatory classes at the moment
To evolve and move out of the traditional form of such things, prep sessions have become a popular commodity in the market for weak souls who like to send comfort to their souls through the words of charlatans and astrologers .. Preparatory sessions since the beginning of the nineties now require one to five people sitting in a circle or in a square manner, depending on the shape of the table on which they sit, and then their hands are intertwined through the door of transfer of spiritual energy between them, often two of these five people are the magician who works, which is also part of the exaggeration .. Assignments with the same naturally evoked the spirit and or extracted some gibberish

muttering to reassure people that he was doing his job to the fullest and beyond .. Silence ..

- O pure soul !. Oh, pure soul .. Oh soul, flan! Are you in the office? Tell us if it's too late? ".. A well-known way of spiritual signals of its existence is to drop something from an antique or hanging picture or extinguish candles, if any, during a session .. Of course, such an atmosphere inspires some fear and anxiety in those present, and they begin to physically tremble and mumble about their religious fortifications.

Between Glory and Shock (The John Edward Experience)
To introduce John Edward, he is a psychic and also one of the best who uses telepathy, this man provoked a wide controversy in 2003-2004 when he conducted a very daring experiment, which consists in reading the thoughts of the audience and those present inside the studio in order to prove it took nearly a quarter of an hour to be correct in what he was doing. He started asking people what they thought about marriage projects and engagements, as well as personal relationships, another aspect of his surprise was that he revealed the origins of each person .. For example, you are the son of flan, flan and flan, and your origins are related with a king flank or queen flank .. And so people kept two mouths open, such things made John more like a prosperous Hollywood and became the most famous of them, because information that poses a danger to humanity should stop him .. His first telepathic experience on air caused controversy in forums and in newspapers because it evoked the spirit of a relative present in the studio.
The truth behind John Edward ...
In fact, the spirit of the person he wanted to evoke was never present and never communicated with John, much to John's surprise and surprise: "How can a capable mediator not summon the soul of an ordinary person?" ... But he downplayed it on the pretext of lack of focus or spirit not passing through the cosmic ocean, and a number of excuses and frivolous arguments.
How does he know the thoughts of those in the studio?
Well that's a good question ... Note, dear reader, that he is asking the audience to bring a family tree with them. As shown in the picture, and based on this, he conducts his research and prepares useful suggestions that help him create an atmosphere of spirituality and trust.

In the end, John Edward, after a big scandal, stopped working, and his last international television appearance was The Oprah Show, where he revealed all the secrets of the profession.

Word from the writer
I would like to start my speech with a quote from the wonderful late Anis Mansour, he said: "If you believe everything you read ... don't read" social media sites and commercial books that promote these things have no purpose other than fame. influence and sometimes make money .. Dear reader, note that no matter how much science developed and what degrees of creative perfection it reached, it stood in confusion and eluded two things .. Death and the soul .. Spirit is the greatest a divine mystery that science has always sought to unravel, but they have failed, and death is a very deep search that opens

thousands of doors and doors for you, and you will not get a convincing answer .. An abstract must be done to hear about these things and live my life.
Are the ghosts real or fake?
We come to the most important phenomenon in the world, which has always occupied public opinion, residents of homes and film producers. Before I start talking about this phenomenon .. I would like to ask: Have you ever heard of a man who was unjustly killed and whose soul went to roam the country in search of a murderer for revenge? ... The answer is no .. Because if this happened, the world would be full of ghosts of killers and killers .. It is important to note that the phenomenon of ghosts and their connection with the Dead is still a mystery, because there is no physical evidence to support the theory of ghosts. as well as their connection to haunted houses, such as this 1936 painting of the ghost of a white lady descending a staircase.

Well, let's talk calmly. it is said that ghosts inhabit cemeteries, poor houses, and generally abandoned places. ...

We all know the US Capitol, or US Capitol, which has been known for ghost stories for centuries, one passer-by saw a black spectrum wandering the corridors, and as he approached it, he discovered that it was the late President Abraham Lincoln while others saw Roosevelt .. Eisenhower .. "This news reinforced the idea that ghosts exist in close proximity to humans, and may have added some sparkle and splendor to it.

Let's start with a test, but I want, dear readers, to focus and use as much as the mind can provide to draw a conclusion, the content of this image .. Well, this picture, some people argue that it is a ghost. what do you see ? Please don't come downstairs, so do your best ..

The image is just a glowing flash in front of the lens during capture .. I look forward to your findings in the comments.

Ghostly things. :

- Haunted houses - A cold spot is usually found in the wall or walls of the House - Nausea and shortness of breath - The disappearance of clothes or small things and their sudden appearance, even if you search for them repeatedly and completely .. - Sounds and smells from unknown sources.
Someone asks a realistic question and says: why do ghosts always appear or are portrayed by people in stories and films as dressed in a white suit? ..

The traditional answer from their books is because they awaken from death with a white shroud and thus form the form we know of them.

One of the most famous stories that stunned and deeply touched people is Sleepy Hollow, the Ghost Rider who was beheaded and returned to take revenge on the filmmakers who brought the legend to life in the movie starring Johnny Depp.

What is the opinion of science on this issue?

Despite the ambiguity surrounding the topic, they came up with almost convincing solutions from them.

- Seeing White Aviaries as a result of chemical reactions, because the houses are located under the sewer system - sewer pipes - and therefore the smells emanating from them have a significant effect on the state of the human brain, causing him visual hallucinations.

- The images are just visual tricks due to the lens not being clean or something else due to exposure to light.

- What you see from the spectrum is just the Genie, because science has proven that the only thing that matters in existence is the Genie .. A good example of this is the phenomenon of noisy ghosts that sneak up at night, throw pots and emit loud noises and then go away.

Word from the writer
I admit that these opinions are not convincing to any of us, but I wrote them to open you to surprise and search for an accurate and healing answer to curiosity. And in the end, dear reader, the world is too big to keep in our eyes or to be trapped in a vicious circle surrounded by these superstitions, these things, the same as the pin-and-pin head from this vast universe, and finally, admit that my article is short because not everything is worth mentioning .. Just remember that Ouija, black magic, witchcraft, quackery and haunted houses are just things that exist in your mind, you scare yourself and laugh at yourself, and always remember the saying of the Prophet Muhammad (peace be upon him): (O son of Adam, if jinn and jinn meet to harm you with something that will not harm you, except what is written by God

Dinosaur park between truth and fiction.

There is always something nostalgic in the past, each era has its own advantages and disadvantages, and despite the technology we live in today, we are, for example, the black and white generation and still enjoy old films more than modern ones, and I mean Arab films, and even if we change our houses to modern times, there is something inside. The past has its own charm.

But what's on your mind if the past goes back a little?

For royal days in Egypt

Back yet ..

Pharaonic times

Back more .. More ..

For the era of the dinosaurs.

Did he hit you like lightning?

Scientists did not see it, but considered it a stunning scientific victory if he succeeded and opened closed doors and that the next ten years would be very different in the world of creatures and embryos and witness a terrible and almost limitless boom.
Long ago, scientists wondered if they could awaken extinct creatures such as the Siberian mammoth, Tasmanian tiger, worm birds and many others, right down to the dinosaurs, the greatest creatures ever created on the face of a simple, and I mean, large in size. , world cinema, who introduced this thing in the series "Dinosaur Park" directed by Steven Spielberg, the best filmmakers based on this film. Then, after a five-year absence, the same director returns to stage a new Miracle, but with new dinosaurs, and again breaks box office records and remains number one on the list of best grossing films for several weeks.

Do you know what's going on?

It is believed that this proves that the audience wants these creatures to return.
And whether we disagree, or agree, there are scientists who support this point of view and agree with it, and instead of thinking about the happiness of mankind and in the treatment of incurable diseases, their brilliant mind is enough about how to return these creatures and did not stop hearing objectors, but completed their journey in pursuit of the deep past, and the first problem of mosquitoes) from dinosaurs when they were its contemporaries. That is, before the plant glue fell on it, it froze inside, and over time, this glue turned into an amber stone.

The second problem that scientists face is to choose an animal that will accept the fetus, any dinosaur embryo is cloned from the amber insect genetic adaptation tape, we are not talking about any animal, but it must be on the genetic Price map read with dinosaurs, or close to her in shape even remotely, and I found scientists found.

Now everything is there and they just have to experiment.
The first step came from John Moore University in Liverpool (UK), where a group of scientists announced that they had finally managed to clone a real dinosaur and give it a name .. Shouldn't they give it a name? How kind these scientists are ... Right? ..

Spot, the little dinosaur, was born at the University Veterinary School and is in good health ..

An important friend of ours (spot) from an ostrich egg, his ostrich was a dinosaur egg in hardness ..

My reader friend doesn't think this is a dream, in fact, a small dinosaur is being raised in Britain right now .. Be careful ..

And scientists have extracted the acid of our friend Spot from a dinosaur fossil named Apatosaurus, and it's worth noting that scientists say that our little friend is completely healthy and does not suffer at all.
We leave our master (the spot) and go to another station in Australia, how beautiful the animals of Australia are, starting with the first Cola bear and Kangaroo and parrots of different shapes and sizes .. But some scientists are not satisfied that, according to them, the living animals became extinct so they said together that they are actually killing kangaroo animals these days, and numbers became an issue and began to compete with humans in their places of control. The implication of this speech is that the Australian mouthpiece says: "We are missing what we have." But some Americans stick their nose into the project until the topic in which they have to be completed to the end, even if they she will have to lay an artificial island near the continent of Australia.

And the bottom line is that these two got DNA from a Tasmanian tiger that was kept in a laboratory, and they transplanted it into a mouse gene, and you say why a rat, although a Tasmanian tiger is about the size of a dog? ... I say, look at the article from the very beginning, and you will understand that size is not important, which is important regarding the genetic band between extinct and neighbors, and scientists say that the experiment worked, but we have not touched or seen any Tasmanian tiger yet, this maybe in the future, just like a trip to the moon.
Come with me now to the new terminal in the book of cloned extinct animals. Let's talk about creatures, as there are many and a lot of scientists about him without stopping, especially since there are many sound samples not, until sperm, that the creature still exists, even its own food .. Everything exists .. And scientists will have nothing cost, except to introduce it into a creature that is closer to it, and become our being again .. And I am talking here about a giant mammoth .. With its thick fur and long fangs, Siberian ice retained most of the carcass of this herbivore, extinct for thousands of years back.

Now, my friend reader, imagine that one day you wake up and, looking out the window, will see a dinosaur walking in your garden of the Caliph, or will the day come when you will go to the circus and instead of an elephant and a lion you will see a mammoth and a Tasmanian tiger? ..

It's absurd that some institutions are desperate to produce such animals in order to create a park dedicated to extinct animals, as if the movie dinosaur park is just a glimpse into the events of the future.

Amazing prophecies: fantasy turned into reality

The dawn of history, the so-called "agrarian revolution", then came the nineteenth century, to deservedly open a new page that changed the features of life in the world, called the "Industrial Revolution".

The same revolution took the newborn "science fiction literature" by the hand, which in turn predicted that if our era is revolutionary in some way, then it is an information and communication revolution.

We are now tracking these sci-fi whispers to see their role in anticipating our current bull and predicting it years before it happens. :

Space stations
The first work on the idea of an artificial device around the Earth is a "moon brick", the story first appeared in 1869 in the pages of the magazine (The Atlantic Monthly) and bore the signature "Edward Hale ", I investigated the details of an artificial creature made of bricks with a diameter of about 200 meters, appointed Astronautics personnel, but accidentally launched the carrier.

Telephone
Work activities are a small tool for communication between ship crews during their offshore field work.

The idea remained stuck in the mind of the viewer of a dreamer named "Martin Cooper", this man in five years will be called the father of the mobile phone, where he turned the idea into a tangible invention in 8 years, and the first call was made on April 3, 1973, but of course , it was not potential or its current size.

The invention of the telephone itself is a whole other story, there are those who pay tribute to Graham Bell and there are those who insist that it is the undisputed Elisha Gray or partner.

It's a very strange coincidence that two of them made the same invention and applied to register it at the same time, if not for the fact that (Bell) a few hours earlier, Gray did not remain silent on his part and filed several lawsuits that ended in Bell's favor.

The first video call in history was made by Bell on April 20, 1964, using technology developed by Bell systems (later Bell Labs).

*****

Arthur Clarke says at this point:

"The meaning and function of science fiction literature is united in developing people's imaginations and giving them the ability to think about the future."

*****

Internet and satellite TV
However, Clarke's career took on various events, where he was inspired by Bell Labs' appeal and led him to write the novel "Terrifying Newborn" six years later, and put forward a future vision of the invention of the Internet, and the most important prophecy of his life, visiting him at the age of twenty-seven , that is, in 1945, all over the world. "

Cited in detail Clark's vision of machines revolving around the Earth, facilitating communication tasks, etc., so far not new, (Edward Hale) pointed out that similarly in (brick of his Moon) that new is Clark's suggestion about the exact shape that governs the orbits of these man-made objects, and he concluded that 35,768 kilometers is the longest distance.

Arthur Clarke did not regret anything, as much as he ignored the registration of ownership of the idea of satellites, his only consolation was that scientists saved his favor, and the orbits were sometimes called Clarke's orbits.

Electronic documents

It is an interactive e-paper capable of constantly changing the content of its face to carry whatever text and images you want.

The first appearance of this idea was in the hands of handsome Tom Cruise , thanks to the events of his film Minority Report, who was behind his camera genius (Steven Silberg) as director.

International companies were the first to come up with this idea, saw in it the complex shape of tablet readers, years later LG and Sony took the lead and introduced their product from this remarkable innovation for the first time.

Credit cards
The novel caused a huge stir at the time, finishing in third place in bestsellers after Uncle Tom's Cabin and Ben- Hur : The Story of Christ.

Surprisingly, the work contained an accurate description of credit cards as we use them today!

The novel tells the story of an American who sleeps in 1887 to wake up in 2000 and find himself in a socialist society, where the government gives every citizen a credit card to buy goods, and the government invests each person in his credit card its share of the country's national product.

The man went further and imagined that these cards can be used between countries.

Blamy imagined not only the shape of credit cards, but also the shape of modern stores, in the plot you will find huge places where goods are stacked, the buyer chooses what he wants, and then pays with a card during checkout, we all seem to us a familiar picture, which doesn't need a genius, but I'll try to remind you:

- By analogy with the author who lived in 1888, she is definitely a genius.

Twenty-one shrapnel women!

Have we ever wondered what happens to children who are being molested? We will have behavioral deviations, or sometimes they will have a phobia of the opposite sex, or ... etc.

But this is a completely, completely different story.

Abused when she was only 3 years old!

Kim Noble was born on November 21, 1960 and raised with her parents and sister in a normal family, her parents worked, so Kim was left with a few nannies and sometimes accompanied by relatives and neighbors, some of whom were more kind and loving, while others enjoyed her, and she was regularly and continuously sexually abused.

Kim was small and helpless, she could neither speak nor compose, and did not even know that the count was wrong! ..

But she's just scared and knows he is hurting her physically and mentally. She had little pain and what she was going through, but she lived through all this pain and exploitation and did not say anything.

So it's time after her second birthday ..

Kim fell like glass falls to the floor, the shards become shards, some of which are small and some more, none of which is ever the same ..

That's twenty pieces!
From one woman to twenty different personalities .. Each character has a different life, a different time and a completely different family situation than Kim, because Kim basically no longer exists !.

No one knew about Kim until she decided to go public for her first television interview at the age of 50, when she lived with her 14-year-old daughter Amy and two dogs.

Some of Kim's characters are male, some are children, and even worse, she didn't know she had a daughter !.

She is the mistress of the house, she is a very talented artist, but she did not have drawing lessons, and her life is a complete mess.

Most of Kim's characters have neither bad memories nor flashbacks , so Kim was protected from what happened to her as a child.

And there is Abby, a young girl desperately looking for love, and Bonnie , a young mother who took her little girl from her and she is looking for her, and Kim often tells her real daughter Amy about her lost child (Skye) and Amy doesn't an answer other than tears to his mother, who does not exist among these characters.

Salome is a devout Catholic, and there is a boy whose name is written only in Latin, Diabal .

Ken's character is gay, Judy is a 12-year-old with an eating disorder just like Kim in her teens, and the controversial Ria character constantly paints disturbing images of children in opposing families being tortured and harassed.

Kim's disease is known as DID and is a rare condition called schizotypal multiple personality disorder.
Kim's characters switch from one to the other without warning.

Kim's case was examined and doctors confirmed that she was in conflict with forces beyond her control!

When Kim seems to wake up from a dream that doesn't take just a few seconds, the character begins to wonder where I am to find myself, either driving a car, in a supermarket, or where else was the character before her.

In Kim's case, there is no character leak, each character works independently and shares the same body, but does not share any experience.

At some point in Kim's treatment, the authorities took Amy as a child and kept her away from her mentally ill mother for 6 months, but psychiatrists helped appeal the decision on the grounds that none of Kim's characters would pose a threat to Amy.

Since then, Kim has received support from psychotherapists and social workers.

Patricia is considered the main character, sells her paintings and has a credit card, only she knows her PIN, and each character has her own accounts and passwords.

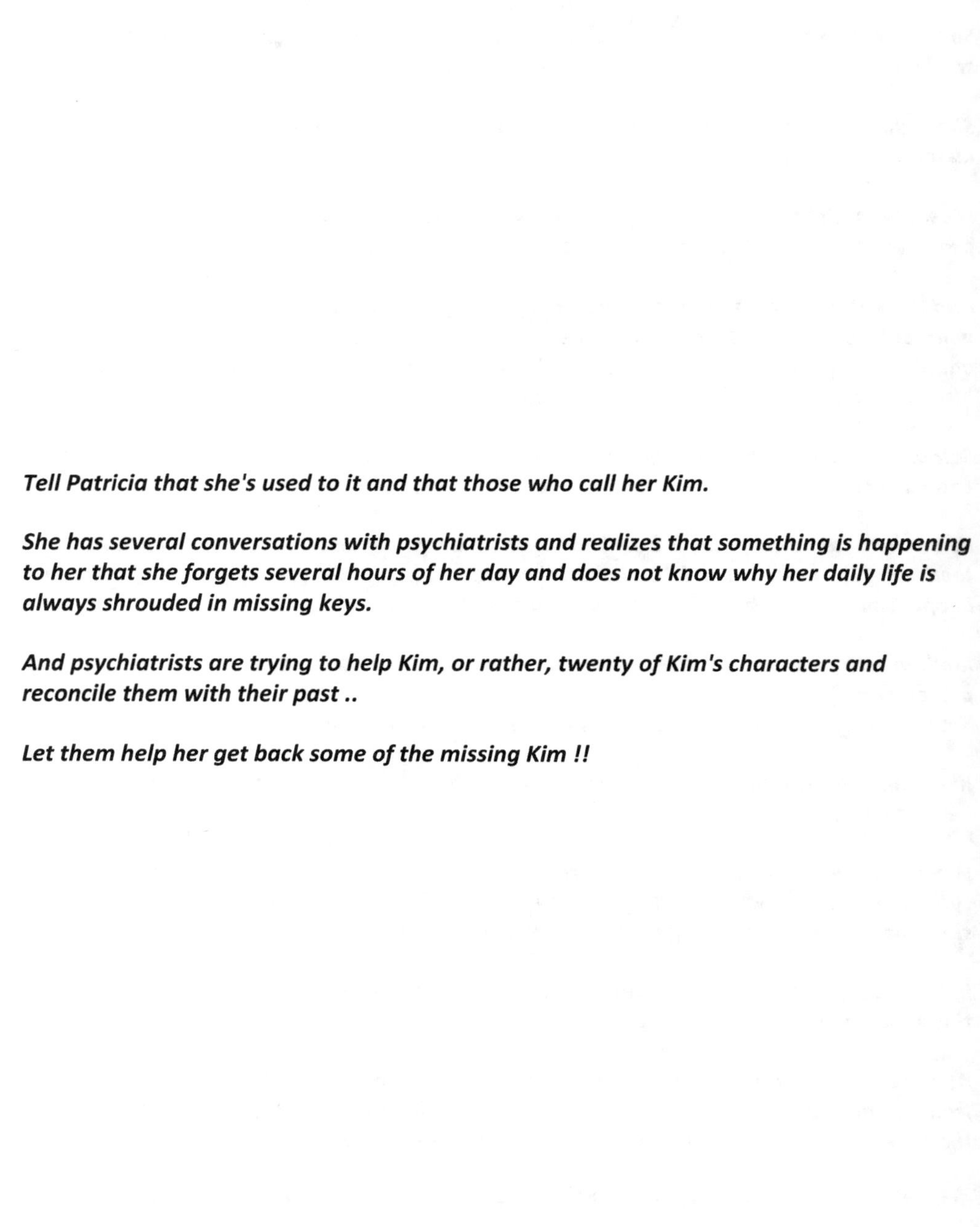

Tell Patricia that she's used to it and that those who call her Kim.

She has several conversations with psychiatrists and realizes that something is happening to her that she forgets several hours of her day and does not know why her daily life is always shrouded in missing keys.

And psychiatrists are trying to help Kim, or rather, twenty of Kim's characters and reconcile them with their past ..

Let them help her get back some of the missing Kim !!

Robots through the ages

*It is a mistake to think that a robot is an idea of the modern era .. The truth is that ..
Because mechanical models have been invented since time immemorial, albeit in primitive
forms .. But the history of these attempts is fascinating and reverent in itself ..*

*From the first century AD it was in full bloom .. - Yes !! You are not mistaken in your
understanding of history .. It was in this era that Heron made his first attempts in
Alexandria, and the result was machines with a simple mechanical movement, driven by
steam ..*
*Then came the day of the Arabs .. Where he (Badi al- Zaman al- Jazari) mixed scientific
and artistic meaning, he created in 1206 a musical orchestra of four mechanical musicians
working independently of each other, and you can push them to create musical
compositions ..*

*And robots continued to work in the hands of artists, giving the Italian genius Leonardo da
Vinci a design for a knight that could perform limited movements ..*

*Until the descendants of the samurai appeared, when Tanaka Hisashiji in 1800 introduced
a series of toys that could serve tea, paint and, among them, shoot arrows.*

*All this was in the era of the first seat .. Then we saw the whole exhibition of robots
presented by Westinghouse between 1930 and 1948 .. Where Elsie and Elmer appeared ,
produced by William Gray Walter , where they could show some simple biological patterns
of behavior ..*

*Then jumping started to speed up when George Devol introduced a machine, installed in
1961, to lift hot pieces of metal from the machine and then fold them.*

*Two years later, Fuji Yuzuki followed with an automated transport robot .. Japan is the
largest consumer of these machines in the manufacturing industry ..*
The Coca group introduced famiols with six electromechanical axles in 1973 ..

*In today's world .. We can see more complex shapes like the P3, which is about 160cm tall
and is capable of jumping, running, opening doors and even playing tennis!*

*Reveng University is also developing a human interaction robot specifically for caring for
children with autism.*

At the same university, Kevin is a researcher specializing in the development of human-machine understanding .. And he has results.

Dazzling by making a microchip, injecting it under the skin of your left hand and creating a kind of communication! So this is what the mind is

The electronic one understands where he is at any time, turns on the laboratory light when he enters, and then greets him!

We can say that the future of this technology promises good prospects for better control of cars, since later you will not need to give voice commands or press buttons.

On the buttons .. A signal from your hand or the sound of a crackling finger is enough ..

Others are also aiming for a more distant horizon so that we connect to the Internet through our thoughts and then broadcast it directly to someone else's nervous system, which is the same idea that Abdul Samad Ghazwani formulated in his story (free cave), where he talked about negative consequences of a world that violates the privacy of the mind and makes it as open as any computer connected to the Internet. His story won the Nabil Farouk Award for Science Fiction.

And if we go back to the first who gave the floor to a robot, then this is the Czech Karl Quebec in 1921, in the play "R. W. R" we are talking about robots made of artificial meat used by the factory as a cheap labor force and different from those who we know as having a high degree of emotion, that is, feeling, dreaming and evil too!

But the truth is that the writer who took the laws of robots from his stories is a child prodigy: Isaac Asimov (1920-1992)
He created the first list of robot rules. :

1.the robot must not harm people, otherwise inaction will lead to harm

2-he must obey any order given by a person, without contradicting the first law

3. On Android, protect your existence until it contradicts the first and second laws

Azimov also predicted these laws in his collection of short stories "I am a Robot", which was turned into a film of the same name.

The title, starring Will Smith, is a robot-hating cop who is investigating a murder, and the suspect is a robot .. In the first scene, the film began with a marble painting on which the Three Laws of Asimov were inscribed ..

In recognition of his work, Honda honored the great writer by naming his latest generation of robots " Asimo. "

And if we look for a photo of him in the Royers archives , we find that he teaches students to dance .. I mean that the robot Asimo is a teacher, not a writer ..

Finally, mention should be made of the American writer Philip K. Dick (1928-1982), aka robot writer who gave us one of the timeless novels about electronic minds, titled Do Robots Dream of Electronic Sheep. This novel is said to have inspired Sony when she created the dog Aibo .)

Science or Destruction? What is CERN?

What is the universe made of? How did he grow up? Do UFOs belong to aliens from another planet? Are they unknown from time to time? Or is it just mass hysteria?

Researchers at CERN are trying to find answers to all these questions and more using the world's largest and most powerful particle accelerator .. But what is CERN?

European Organization for Nuclear Research is the abbreviation for the European Organization for Nuclear Research, known as the Cernne Collider or Large Hadron Collider ..

His goal: to understand the structure of the Universe, find a way to understand its constituent particles, what their types are, how they are affected, and deeply understand the laws of nature, the genesis of space, its state and ultimate fate.

Founded in 1954, it is one of the first European joint ventures between 21 countries

It is a facility where physicists and engineers investigate the reality of the basic structure of the universe, and they can use the most sophisticated scientific tools to study the basic components: particles. Where are these

A huge machine collides subatomic particles in an accelerator circuit at a speed close to the speed of light .. This process provides scientists with the key to understanding how molecules interact, and allows them to explore the fundamental laws of nature ..
This accelerator consists of a particle collider and collision detectors.

The collider is located inside an underground tunnel near the Franco-Swiss border and has a depth of 575 feet, which is equivalent to about 175 meters ... The cost of this project has reached more than $ 9 billion ..

One of the most famous discoveries: the discovery of the "God particle" in 2012, which is a very important central particle, is

A new understanding of the structure of matter, how the Universe arose and the great theory that governs it ... And scientists believe that this particular particle is responsible for acquiring water for it ..

This project is an important step on the path of the imam in the field of science and knowledge, but there are many scientists and priests who warned about this experiment and demanded that it be stopped.

So is CERN a threat to peace?

This collider can bring unexpected disasters and dangers to this universe .. And the scientist Stephen Hawking is one of those who

They oppose this project, as stated in the preface to Starmus : this is a collection of lectures

Submitted by Scientists .. In it he says: ((A newly discovered God particle could destroy the Universe as it can become unstable if exposed to very high energy levels, which suggests *the possibility of a catastrophe leading to the collapse of the Universe, we are at a certain risk from -for this))*

And scientist Hawking is not the only voice condemning these experiments, there are others who agree with him: astrophysicist Neil De Grasse Tyson , who said in a conversation on American radio: "The experiment could lead to the explosion of the universe."))

German professor and scientist Otto Rosler also filed a lawsuit against the project leaders in 2008, arguing that the collider could produce a black hole that could swallow the entire earth !!)

But the lawsuit was rejected, and the collider continues its work to this day .. And some scientists believe that the project will create a powerful explosion that will destroy life and turn the universe into a vacuum!
One of the most unusual and exciting things is the CERN Collider logo , which exudes a lot of pessimism and conveys

Fear lives in the hearts of many because of its satanic symbols!

The image here depicts a logo with the number six BIS three times 666, which is a symbol of the Beast or number

The devil .. This sign caused alarm among the general public and the clergy, who considered it a satanic experiment, the purpose of which was to destroy all existence and the universe ..

In addition, a strange statue was chosen to represent the project, as shown in the figure: a statue of a Hindu goddess named Shiva, the God of Doom - destruction-destruction, or destroyer of the world, as it symbolizes cruelty and doom and embodies the power that works to destroy the universe with all its living creatures, even the planets themselves!

Science or Destruction?
What is CERN?
The statue representing the project is a Hindu goddess called the goddess of fate
And the question that has puzzled minds is how can a scientific organization accept the choice of such symbols and logos? Is all this a soft declaration of the imminent apocalypse and the seizure of power by Satan?

Confusing questions are still looking for answers!

Watch your hand! it can kill you!

Maybe I once ran into someone who behaved strangely and freely in one of his hands, can be annoying and sometimes embarrassing for Wales Docking to one of his hands under the pretext that they are behaving without any interference on his part. You may even find that he is holding his other hand and yelling "stop." You may think that he is crazy or sick himself, imagine that they want to kill him around him or that his arm has been bewitched. But the truth is, never.

Alien Hand Syndrome)
It is a chronic condition caused by traumatic brain injury due to a stroke or brain surgery, which causes the patient to become unable to control their arm and causes the arm to make movements beyond the patient's control, and may even embarrass him in front of

people by making shameful movements such as touching sensitive organs, or being a threat to the patient himself, for example, choking him, pulling his hair or forcibly stuffing food into his mouth, prompting him to scream desperately in an attempt to control them without benefit.

Pathological conditions
In 1908, German physician Kurt Goldstein published a report on a forty-year-old woman who had a strange condition: after a stroke, she suffered from cerebral palsy and eventually recovered from it, but the woman was horrified to see her left arm move without feeling it. and not controlling. it got to the point that her hand movements became dangerous, according to the doctor's description of her condition. the woman's psychological state deteriorated as a result of these involuntary movements, which became terrifying over time until she died a few years later after her hand movements worsened.

Many people have the same condition after brain surgery or suffered brain damage that caused them to lose control of their arm, one published report cites the case of a 68-year-old woman who claimed a ghost pulled her arm but warned her medical staff to approach her so as not to suffer from her violent hand movements, which almost practically strangled her ..

Is this story true? I don't know, but there are many videos on YouTube depicting some patients suffering from this syndrome and documenting what happened to them suffering from their hand.

Alien Hand Syndrome, after all. do not treat her, all he knew was that the damage to the corpus callosum affected the brain as a result of surgery or a tumor associated with the brain.

Mysterious places on the planet

Belt of Silence or Zone of Silence)
Who among us, dear reader, does not seek complete tranquility away from the surrounding noise? And who doesn't want to be alone where no one can get? There is a class of people who prefer the above, including myself .. The Belt of Silence is the right tourist destination for this category of people, there all communications are interrupted by phones or wireless devices and others, but let me disappoint you dear ones, because it is difficult to get there. even to the drug dealers themselves, for it is known that they dare to enter every place, even they do not use this place, although it is the best.

In the Mapemi deserts in northern Mexico there is a zone of silence, where complete silence reigns, this place has witnessed many strange phenomena and accidents, for example, all communications are interrupted in this place, neither the phone receives its signal there, nor the Compass can determine it destination, and even JPS doesn't send or receive its signals ...

Some strange things that happen in this region

After the American missile crash in the area, followed by the Saturn missile crash, the reputation of the area touched the target of every explorer and scientist or even adventurer, they went there to find answers to their questions, but when they arrived, then they found only questions above what they have, where they noticed the strange we are talking about one or two animals, but there are many of them. They also noticed that when the turtles are placed in this place, they turn on their back, and the reason for this is not yet known, and on one of the expeditions one of the explorers was accompanied by his own dog, and when they arrived at this place, the dog barked strangely and creepy, as if she was afraid of something, and then the dog ran out of this place and did not find her .. During one of their trips to this place in 1985, they had to stay there for a week, but stayed only three days and ran away, and the reason is that people who have been on the trip show strange signs such as hallucinations and fatigue, and the night sky in this place is scary due to the sight of celestial bodies.

Some theories to explain strange phenomena
Interpretations of what is happening in this mysterious place range from people's beliefs, which are often based on the interweaving of their imaginations, to the opinions and theories of scientists trying to find a logical explanation for everything that happens.

There is a theory that this place is surrounded by powerful magnetic energy, which allows aliens to open the Wormwood passages through the universes, from which they pass into our world. The most likely theory is that this place emits radiation as a result of the fall of many celestial bodies on it , which leads to the death of animals and the interruption of all communications ..

Hell Gate
Somewhere in Siberia, scientist Azakov and his team of scientists sent by the government of the Soviet Union dug a large hole in the ground for research, and when they reached a depth of 14 thousand meters (14 km) below the surface of the Earth, Azakov decided to carry out some tests and experiments; since no one had ever reached this distance, he ordered his staff to drop special devices down into the depths of the crater, and here's a surprise !! One of the crew members heard a man screaming in the hole! .. But some of them at first glance thought that this was the sound of the equipment used in the exercises, " Azakov " ordered them to turn off all equipment and be silent, and after that they returned listening to this sound, but !! .. This time it was not the voice of the person who was being tortured, but the men and women who were tortured in the dungeon, as if they were evil people who were tortured in hell ..

Even though they heard these voices, they decided to continue their work, and the deeper they went into the ground, the more intense the voices were until the decision was made in 1994 to terminate the project.

Sounds that were recorded by scientists
Of course, dear reader, the only evidence to support this statement (that Hell is raging in the earth, where people are tormented by evil) is the audio recording of these sounds, and you know that everyone can manipulate the recordings and add sounds and effects to them, and this is what, that the American tabloid newspapers said that this pit-hole full of

natural, that the tortured voices of the people inside the pit are just a clip from the soundtrack of the film. Blood of the Baron, which was released in 1972, has been reworked to sound like people torturing a hole .. Trinity network (TBN) has stated that this hole does exist and that this is the hell it was founded in. on some lines from the Bible, and (Ogg Rendalen) the Norwegian teacher decided to play his game on people, where he first refuted this theory and said that it was just a burning hole, but he did not broadcast, but changed his mind and decided that the crater flies in the sky of the Soviet Union ..

What do you think, dear reader?

Satan's sea
Japan, a country where technological development is great, and where its population has exceeded for centuries in all matters both scientifically and practically, and of course Japan will not lose sight of the development also due to mysterious incidents when they have there is the so-called Devil's Sea or Dragon's Sea, which is a vast body of water surrounding the Japanese island of Miyake . This place has far surpassed the Bermuda Triangle in terms of mysterious disappearances, for which there is no logical explanation yet, and unlike the Bermuda Triangle, the disappearances in the sea of Satan continue to this day .. Wow, even in such things, they do not allow anyone to outwit themselves! ! In 1987, the Philippine ship Carlota entered the sea of the devil during its Trans-Pacific voyage, and suddenly and without warning caught fire in the front of the ship, not knowing why, and knowing that the ship did not carry flammable materials on board, the crew members rushed to extinguish the flames but they still noticed that a fire had arisen. The crew members continued to fight the fire and put it out from the front of the ship and at times behind it, and after they fled and fled in the direction where they successfully extinguished the fire, but the ship was badly damaged and returned to the Philippines, the ship is still still anchored in the Philippine port to this day and has.

The Japanese ship Kayo once sailed into the devil's sea to study volcanic activity in its 31-man hall, and of course it mysteriously disappeared, but before that it sent one distress signal ..

Three naval submarines, two of them with nuclear warheads on board, also disappeared into the sea of the devil.

During World War II, four warships disappeared under mysterious circumstances, and despite efforts to find them, they came to nothing ..

Between 1952-1954, five Japanese warships mysteriously disappeared, leaving the Japanese government with about 700 passengers on board.

In addition, in 1949-1953 seven small fishing boats disappeared there for unknown reasons ..

Almost every year there are frequent cases of disappearance of fishing boats in this place, and their number can reach hundreds. But the incident that angered the Japanese government was its preparation and funding of a research vessel belonging to a huge project to explore and study the Sea of Satan, and on more than 100 Japanese scientists were on board, and it is worth mentioning that the vessel was equipped with all the modern equipment for communication and research and was provided with everything necessary. Here the Japanese government has declared the severity of this area and developed within the restricted areas, which gives anyone approaching them an education around the world.

Like the Bermuda Triangle, several theories have been created about what happens inside the sea of the devil, some attribute these oddities to the otherworldly and alien, while some provide logical analysis and explanations, but, unfortunately, without convincing evidence to support them ..

The destruction accepted was one of the theories accepted by scientists, they said that mysterious accidents and disappearances were caused by two reasons, the first of which was war, as the region was going through a period of wars at that time and therefore the sides of the war may have caused the disappearance of ships and planes .. The second reason is piracy, where they put the blame on the back of the pirates, they put the possibility that pirates were stealing their ships, but do pirates also steal planes? Maybe, who knows ..

Human error was also among the theories, as people are mistaken in most of their questions, they said that the disappearances were caused by human error .. But are all these accidents the result of human error?

The Japanese government has also stated that this area has gravitational disturbances and that the gravity at this location is unbalanced and therefore pulls anything that passes from its range into depth.

Some explanations were unnatural, they said that aliens were the cause of what happened, and others said that the devil's bastion was in that sea ... As you can see, dear reader, this is the same scenario as the Bermuda Triangle, but with a different pattern where you find these incidents, you find these completely irrational explanations, what do you think?

Secrets of the Universe

There are many mysteries in this vast universe that have not yet been discovered and mysteries that remain unknown to this day, some considered them pure myths and nonsense inspired by science fiction such as black and white holes and worms, until they proved their existence, and we're going to understand them now more than at any other time, and God willing in that.

Black holes :
We have all heard about black holes, but in order to understand what a black hole is, we need to correct some ideas about it .. Firstly, this is not a hole, not a hole, not a flat object, but an opaque spherical object, like most objects in space, which revolve around themselves under the influence of their high gravity ..

How a black hole is born:
For a black hole to appear in a star, certain conditions must be met. First, a star must be at least twenty-five times larger than our Sun, and at best there must be only one star out of every thousand stars of this magnitude, and scientists have calculated that there are more than 100 million stars of this magnitude in the Milky Way, and these stars are composed mainly of hydrogen and a small amount of helium .. Gaseous hydrogen and helium begin to condense and rotate around each other until it presses on the core, and the heat increases until the hydrogen merges and gaseous helium is formed .. This process is considered the fuel of the Sun until this fuel runs out and the Sun resorts to burning helium and burning all the remaining elements such as iron and carbon. until the star collapses and leads to this Supernova phenomenon, which is an explosion of the shell of a star and the release of all energy and gases into space ..

First white dwarf:
It is a small star, but its density is very high, 100 times that of large stars, it is initially bright and white, and then its color gradually fades, spreading gravitational waves instead of energy waves.

Second, a neutron star. :

It is the remnant of a large star, the size of which is very small. The gases begin to contract on their own, and the density of the star increases until this process produces neutrons, which is why scientists called it a neutron star.

III black hole:
The remaining matter of the star collapses on itself, forming a small but very large mass (black hole), and if this mass revolves around itself, then the black hole has a spherical shape, but in rare cases the mass does not revolve around itself, so this does not lead to completely spherical shape .. When mass begins to revolve around itself, it devours stars and planets nearby and attracts photons of light and gases and makes them revolve around itself, forming a cloud of light, colors and gases, hence the name of the black hole, because it looks like black a hole surrounded by clouds of gas and flowers .. As soon as the hole begins to swallow until it grows in size, material and gases come out and its gravity makes it revolve around it .. Because of this process, we cannot follow behind the largest black hole found in our galaxy, because it is the center of a large amount of gases and materials, which makes it difficult to observe, but in 2017 we will we can see how, according to the European American project (Event Horizon Telescope), we will connect and assemble more than seven huge telescopes, set up the Very Large Telescope so that we can see black holes ..

What makes a black hole a mystery?
The point of no return or "event horizon" is a region that extends from the center of the black hole and reaches a length of millions of light years .. When you enter its range, the black hole's gravity can pull you into the hole, but what happens inside? This is what scientists are trying to uncover by understanding the theories put forward to explain what goes on inside the hole, but there is no practical way to prove it or categorically deny it.

First, the theory of space-time. :
The theory is that if you spend just one minute at the edge of the event horizon, it will be over 1000 years on Earth, and if someone is watching you from afar, you will appear to them as two copies, a copy that moves very, very slowly. , and a copy, which is completely stopped and fixed .. And all this is due to the deformation of space-time near the black hole, that is, space and time do not move normally ..

Cosmic string theory:
The theory is that after the Big Bang, cosmic strings were formed, each of which contained a different space inside, particles moving at the speed of light were very small and accurate and had a length of more than a million light years .. If one of these tendons is near the horizon of events is intertwined with another chord, then this can lead to a closed zone of space-time, that is, a person can enter this zone and travel through the past and the future, calculating the deformation of space-time with mathematical equations .. Of course, scientists could not prove the correctness or inconsistency this theory ..

Gravitational Singularity Theory:

The theory is that at the bottom of the black hole there are singular particles, the size of which is very small, that is, the smallest part of the universe that you cannot see, even if it is billions of times larger, but their mass is heavier than anything The Universe, as in the Big Bang theory, that is, with the explosion of one gravitational molecule.

Firewall theory:
The theory is that if you crossed the event horizon and entered a black hole, you would end up where you weren't even left with anything due to the firewall at the edge of the black hole, because the material heats up significantly inside the black hole component. which can be compared to a wall of firearms ..

Stretch theory:

If one or any object can cross the event horizon, it will expand by gravity and increase its length until it breaks up into small atoms and is pulled into a hole, and scientists called this theory the theory of "spaghetti" because the human body can transform and expand like spaghetti spaghetti due to gravity .. Believe it or not, this theory is closest to scientists ..
And of course, there are many other theories, such as the multiverse and women .. That is, if a man with a ship enters a black hole, he will copy more than one copy of it, that is, a man consists of electrons, nuclei, etc. And the whole this material can be copied as soon as it crosses a black hole to the other side, like the multiple universes theory .. There are several versions of you and the world available through black holes, and of course, these are the weakest theories that scientists have almost come across to refute with the help of some laws of physics and many other theories.

The end of the black hole. :
Everything in the universe has an inevitable end, like planets, stars and, of course, black holes .. Scientist Stephen Hawking discovered that black holes let in radiation that he did not know about, but he called it " Hawking radiation ", and that once it will end, the black hole will collapse .. But scientists believe it could take so long that the entire universe will end and the black holes will eventually be left alone, and this could happen in more than hundreds of billions of years.

White holes:
If there is anything more mysterious than black holes, then they are definitely white holes .. What are white holes made of? It is a natural anti-black hole, so that a black hole absorbs material and destroys it and erases it from existence, while a white hole happens in it exactly the opposite, where it produces material from nowhere and exits from it into wide space. and the meaning of its name "white hole" is not that its color is white, but all. Is this proven? Okay .. White hole only exists in theory, as scientists have not been able to prove it as convincingly as they proved the existence of a black hole ..

What makes the White Hole mysterious?
The fact is that the deformation of space-time in a White hole is so many times greater than in a black hole that it will pass hundreds of millions of years outside the White Hole, and only a few minutes will pass inside it, and besides, its event horizon throws matter out and does not absorb it exactly the opposite of a black hole. After all, the hole remains just a theory on paper that hasn't been proven, but before it, a black hole was just theories

and speech on paper until it was proven to exist .. If it were proven, it would confirm Schwarzschild's theory of wormholes. which requires a black hole absorbing matter at one end and a white hole giving matter at the other.

Wormhole theory:
In theory they are two-way paths that can travel from one end of Time to the other, or from one end of the Universe to the other end of the time-space zone just moments later, but no equation exists until it proves the possibility of a wormhole; they are mathematically impossible inside black holes in the event horizon .. But in trying to prove the existence of a wormhole, scientists tried to prove the existence of a white hole, because this is the only way a wormhole can arise by connecting a black hole to another white one .. There are some scientists who have come up with several plausible theories ..

Alien Matter Theory:
That if a wormhole is found, then it can be forced to settle with the help of a foreign substance .. This bubble is important for preventing the wormhole from collapsing on itself, for example, forcing the traveler to safely cross to the other side and preventing both ends of the wormhole from closing on themselves and completely collapsing .. But they could not find or invent such material .. There was a team of scientists trying to reinvent the wormhole, but after years of effort, they found that they even managed to invent a material so strange that they would have to make every molecule with microscopic money by going into a nuclear laboratory and trying find elements that are not in the periodic table of the fund .. And this means that the task is practically impossible ..

Definition of some terms
Space-time:

Why do we so often refer to the term "spacetime" when talking about black and white holes? When we talk about them, we must consider the Universe as four dimensions: length, width and height .. These are the basic dimensions, but you must add a fourth dimension-time .. Hence the term space-time (space and time) found a simple example: goes is time on the sun at the same pace as time on earth? - Of course not .. Time goes much faster in the sun than on Earth .. Well, if you take an atomic clock into space next to a black hole, and others on Earth, you will find that a clock on the event horizon, reading it, will be completely different .. That is, it will be considered for many years .. The earth will be the same, so the time warp is a very dangerous thing, and no one can get close to it or figure out how to control it, because it will allow us to travel in time and return to the past, and none of this will happen even in a thousand years ..

Gravitational waves:

This is the theory that Einstein predicted 100 years ago, which is waves caused by the collision or explosion of stars or black holes .. Only now that gravitational waves emanating from the merger of two black holes were discovered a billion and 300 million years ago, that is how many it took a wave of time to reach Earth and be seen ..

The Secret of the Poisonous Woman

- You're finished ..

Gloria Ramirez said this after she put the finishing touches on a chocolate bar she was decorating while her two children were flying around her, 12-year-old Evelyn and 9-year-old Angel.

- Come on, Mom .. Start cutting .. Come on.

- Good, dear children.

The hand holding the knife even cut the killer, but she suddenly felt dizzy and could no longer stand, and fell to the ground amid the screams of her children, who did not understand the meaning of what they had received !!

On the evening of February 12, 1994, the General Hospital in Riverside County , California was quiet, disturbed only by the sounds of doctors and nurses moving between wards.

Nurse Susan Kane entered the nurses' room and, yawning, sat on one of the benches, her fellow resident Julie Gorzynski looked at her and said:

- What's wrong with yawning! It's still early, it's already two o'clock and only a quarter.

- Oh .. This is my habit. Whenever I have a shift, I fall asleep early, maybe it's a psychological state.

"You can take a little nap, it's quiet here, and I don't think anything will happen tonight."

Susan came back, she, then got up and said: - No, I will not leave people, I will go get myself a cup of coffee, maybe I will return my activity .. Can I bring you one?

- No, honey .. I have no desire.

Susan got out and went to get coffee, but on the way back noticed a commotion in the hospital and rushed to find out what had happened and almost made it to the nurses' room until she came out with Julie's face and almost hit her ..

- What happened, Julie ? What's that noise?

- We have a state of emergency .. This is me in the emergency room took

She hurried away from Susan, who immediately recovered upon hearing of the emergency, put down her coffee and rushed to catch up with her colleague.

- My aunt .. Will mom be all right?

- Yes, Darling .. Yes.

This is how Maggie Ramirez answered her niece's question, although she knew that her sister was complaining about cancer.
In the emergency room, a team of doctors circled around Gloria Ramirez , whose health was very weak, her heart was beating weak, she complained of shortness of breath, barely answered any questions with vague answers, and in an attempt to save her, the doctors injected her with sedatives and tension, and into a vein injected muscle relaxant , and when noticed.

At that moment, the paramedics saw a shiny, oily substance covering her body and smelling of garlic. Nurse Susan Kane decided to take blood from her hand to find out why this substance was formed. as she did so, she noticed the smell of chemicals coming from the syringe containing the blood. to her surprise, she gave it to psychotherapist Maureen Welch, who brought her nose closer to the syringe and smelled ammonia. she, in turn, passed it on to her colleague Julia Gorzynski , who noticed colored particles floating in Gloria's blood.

But she didn't finish the sentence because Susan suddenly felt dizzy and passed out amidst general numbness.

Susan was taken away from the emergency room and the medics continued their work, but after a while Julie felt dizzy and nauseous and left the intensive care unit, staggering, and as soon as she entered the nurse's room, she also fell, and her colleagues who saw her, they said that she was breathing intermittently and with difficulty !! Maureen Welch also fell, who later said: "I felt nothing but a fall, and when I woke up, I found that I could not control the movements of my limbs."
After what happened to Maureen Welch, several workers and a paramedic complained of the same symptoms, prompting the hospital director to declare an internal emergency and order the evacuation of the intensive care unit and the patients to the hospital parking lot. Only a small team remained in the emergency room. doctors, who were struggling to save Gloria's life, gave her electric shocks, but the pressure continued to fall and her heartbeat began to weaken until she completely stopped and died at the age of thirty-one.

Outside, in the parking lot, hospital workers were trying to rescue their colleagues ..
Susan Kane was convulsing and kicking with her legs and arms, her irritated face seemed to be burning, Julie Gorzynski still didn't come to her senses, and another nurse, Sally Balderas , began to vomit and felt burning sensation in the skin.

From eight quarters to fifty-eight minutes (time of death of Gloria) .. Those 35 minutes turned the hospital upside down, and what happened for them, incredibly, fell on 23 emergency department workers .. The cause is unknown, but the condition itself was strange !! What is it that has spread between them and made them all suffer almost the same symptoms? ¡To find out what happened, they contacted the California Department of Health, which in turn sent two scientists to find out what happened, while Gloria's body was stored and isolated in an airtight coffin, angering her parents who wanted to know the cause of death. and the circumstances of the case ..

***

Ministry of Health envoys Dr. Anna Maria Osorio and Dr. Christine Waller arrived at Riverside Hospital on the same night as the incident and began an investigation.

They interviewed victims in the emergency room, applied a standardized questionnaire and came up with the following results .. Almost all of the infected had the same symptoms of loss of consciousness, shortness of breath and muscle spasms, and they also noticed that people who had direct contact with Gloria had severe condition, and women were more amazed than men! Those who did not have dinner suffered more than those whose stomachs were full, but nevertheless the results of tests and analyzes carried out on them were normal !!

At 11 p.m. the same day, a hazardous materials team arrived to immediately search the emergency room for any toxic chemicals that might be causing the incident, including hydrogen sulfide gas, also known as "sewer gas", which smells like rotten eggs and the concentrated amount is enough to kill after inhalation, but they found nothing!

The Riverside forensic medical examiner , which was faced with the not so easy task of opening the body of Gloria, was also called to investigate the possibility of the presence of toxic substances in her body that poisoned the doctors who rushed to save her life ..
The autopsy of Gloria was carried out after taking all the necessary precautions, for this a special room was built. the doctors were wearing protective vests to protect them from potential toxic emissions, which took several days. The forensic science bureau remained silent and did not provide any details of the autopsy. they decided to use the Medical Science Center at Lawrence Livermore's laboratory east of San Francisco. the center formally intervened in the investigation on March 25 at the request of the forensic science bureau , which contacted the center's deputy director, Mr. Pat Grant. thus, the autopsy samples were sent to the laboratory.

The center's plan of action was to analyze organic and inorganic compounds in the blood, bile and tissues of Gloria's organs, including the heart, lungs, brain and liver, and they investigated the possibility and presence of toxic gases generated in her body and focused on the bile in liver, which is the center of toxins, but still found only air components. And then Brian Anderson, who is the director of the Center, analyzed the identity of the vehicles and drugs that Ramirez took before she died, hoping to come to a conclusion, and the research and analysis took a long time.

On April 12, Anderson arrived in Riverside , showed the coroner the results of his investigation, and said that he had found something similar to poison: Ramirez seems to have taken a lot of codeine, which damages the liver when taken in large and continuous doses. he mentioned the presence of other substances that could cause the smell of ammonia. however, despite these results and these studies, he did not find anything convincing to cause the death of Gloria and the poisoning of the medical personnel who helped her. it made Dr. Anderson almost insane after all that long search.

It was then that the coroner's office saw that the investigation had come to an end and there was no point in making another attempt, so she issued permission for Gloria to be buried two months after her death.
On the morning of Wednesday April 21, 1994, Mrs. Gloria Ramirez was buried at Olive Wood Memorial Park in Riverside , leaving behind two children from a previous marriage. on that gloomy morning, she mixed feelings of sadness, sorrow, resentment and anger in the souls of her relatives and friends, they were outraged by the shameful treatment that was done to her body after death, as well as the fact that
Gloria Ramirez was a kind and friendly woman, who always smiles and brings joy and happiness to everyone. The irony is that, despite all these good qualities, after her death she became known as the poisonous woman or the absolute woman of poisonous fumes !!

On April 29, a press conference was held at which Riverside medical examiner Scottie Hill announced that Ramirez had died of kidney failure caused by cervical cancer she suffered from. Hill said the investigation into her death was complete. Regarding what happened to the hospital staff and the possible connection to Ramirez's death , Hill said that extensive toxicology studies have not identified any external toxicants that could have contributed to her death.

This finding, along with the autopsy and investigation of the hazardous materials team and the questionnaire that was applied to the workers, prompted the Ministry of Health to release its official report in September, which said the outbreak among workers in the emergency department was caused by social illness caused by stress. and anxiety, in other words

Julia Gorzynski denied that she suffered from mass hysteria and filed a lawsuit against Riverside Hospital and the Forensic Bureau, seeking compensation for damages caused to her. - She was also against the state's report of the incident and wanted a trusted authority to re-investigate what happened and she couldn't find anything better than Livermore Laboratory, so she contacted him and sent him all the investigation details and autopsy fragments and convinced him re-investigate and investigate Gloria's case and what happened to her.
Therefore, Anderson asked his deputy, Pat Gratt, to conduct a second search .. Grant went through all the files in the case file, looked through the autopsy results and all his money related to this case, and decided to take another turn in search .. After a long series of research, experiments and analyzes, lasted until early October, the Livermore laboratory suggested that Ramirez rubbed her body with dimethyl sulfoxide (DMSO) as a home remedy for her pain, this substance quickly penetrates the skin and, upon contact with it, appears on the surface of the tongue, giving the taste of garlic, and is sold at Chloe's stores .

They also said that the oxygen that the medical team tried to supply in combination with DMSO and dimethyl sulfone forms DMSO2, which crystallizes at room temperature, and these crystals have been observed in Gloria's blood.

The electric shock she received converted DMSO2 into dimethyl sulfate DMSO4, a toxic and powerful gas that causes symptoms similar to those experienced by emergency room staff.

In this regard, organic chemistry scientist at the University of Wisconsin-Madison, Hans Reich, says he suspects DMSO fission in a relatively cold environment, such as the human body, because the substance needs more temperature to fission, and the symptoms that affect emergency room members relief, not like the symptoms that can be for the eyes and crying, since the symptoms of dmso4 gas need several hours to appear, not minutes, as in the emergency department.

There is also Stanley Jacob , a scientist and research chemist at Oregon State University of Health Sciences in Portland , who has a lot of research on DMSO, who was vehemently against the Livermore test results, and he said it was absurd that this substance was fused at a cold temperature. and stated that he received many calls from women who used the substance to treat interstitial cystitis and were in a panic after this report, but he reassured them by saying that the theory of dimethyl sulfoxide is chemically impossible.

But on the other hand, there are scientists who support this theory and say that it is impossible to say what actually happened, so this theory remains the closest scenario to the truth.
And we cannot conclude the Gloria Ramirez case without mentioning the possibility that it was hidden and not discussed, namely that the hospital staff may have mistakenly confused the use of drugs, using urine mixed with bleach that was in a nearby basin in the emergency room, which led to the formation of toxic chloramine gas and spread into the atmosphere, causing symptoms that happened to paramedics, but no one cared about this theory.

These theories and research carried out in connection with Gloria's death make it clear to us that the human body is a stock of several toxins ready to leave if the atmosphere is right! We often take painkillers and pain relievers and treat them, but all of these drugs are actually toxins, toxins that we swallow with full consciousness and consciousness in search of healing, we take them and forget, but our bodies do not forget and keep their effect ...

Gloria Ramirez led a quiet and monotonous life away from public scrutiny, but her death permanently made her name in the headlines of newspapers, magazines and mainstream news, and several documentaries and programs were shot about the circumstances and circumstances of her death, and in fact the name of the poisonous the lady she was given might have been like it happened at Riverside Hospital on the evening of February 12, 1994? ؟

No one knows for sure, and this case will remain a mystery, like much in this life.

Hysteria.

Hysteria in a nutshell is a flow of uncontrollable emotions and excitement, as well as an illness with the consequences of psychological, physical, who left someone blind, images, or paralysis, or a health emergency that has no grounds for virtual or mystical, any or medical transmissions proving that this person is completely healthy physically, but suffers from the consequences of the disease and severe repetitive ones. This is a short scientific definition of the word hysteria, but people do not use this word only to refer to mental illness, but its content even extends to ordinary people, when they are freed from the bridle of their emotions and emotions, they say, for example, that people had a fit of crying or laughing or hysterical anger.

Hysteria can affect not only one person, but also spread to a group of people in the form of an epidemic, then known as mass hysteria, often as a reaction to feelings of fear, anxiety or threat. In this short article, we'll look at some of the most famous and bizarre cases of mass hysteria, recorded in antiquity or in recent times.

Dance hysteria epidemic
Historians of medieval Europe mention several cases of the so-called "epidemic dance", the most famous and well-documented of which are those that occurred in 1518. The whole story begins with a woman named Mrs. Trovia who suddenly dances in the street in the city of Strasbourg in Alsace in northern France, and soon other people join her.

The strangest thing about this dance is that it was uninvited and unnecessary, without vocals and without music .. However, every hour more and more people joined Miss Truvia in her frantic dance, until there was a great commotion in the city.

It is amazing that these men and women continued to dance without pause for hours, day and night, and for days on end .. So much so that some of them passed out from exhaustion and fatigue, and some died from fatal heart attacks.

Mr. Truffaut stopped after six full days and continued to dance, but this did not mean the end of the epidemic, as now hundreds of people were dancing like crazy in the streets of the city.

The priests and people from the "decision and contract" in the city talked for a long time with the dancers, tried to persuade them to stop, sometimes in good faith, and greeted them with force, but to no avail, as the dance went on non-stop.

Eyewitnesses who survived and survived this extraordinary event spoke of very strange things, including that these dancers were in a subconscious state and that their emotions and movements were involuntary. And for some reason, they cannot look at Red and start acting violently and aggressively if they see it.

And of course, since we are talking here about events that took place in the sixteenth century, where people's faith in the visual was strong and ingrained, this wonderful hysterical dance was associated with magic and black curses. But the city council did not create seemingly sane people. a more logical explanation, so they decided to seek advice from some doctors, twisting the doctors 'opinion that what is happening does not mark a high proportion of bad blood in the dancers' bodies, and bloodletting is a method used at the time of bad blood development, but it was a funny camera for those dancers who jump right and left, I suggest that the doctors instead leave the dancers in their path until

they burn her with bad blood in a continuous movement. And exactly where the streets were cleared for the dancers, a large wooden dance platform was built, and musicians from all over the city gathered to play for them non-stop when the other team got tired .. And now it turned into a big party !.
As the days began, the dancers fell one by one, either due to extreme exhaustion or death from dehydration and heart attacks. By the end of the month, he was gone.

Hundreds of years have passed since this event was documented in church and city archives and supported by the testimony of historians .. However, the debate as to why this happened continues. Today's doctors disagree with the old Bad Blood theory and propose another theory based on the fact that what happened was caused by a fungus growing on rotting wheat, perhaps there was a lot of rotting wheat in the city at the time due to some environmental conditions or poor storage, so the proportion of the fungi multiplied to a high level, causing a kind of mass poisoning that affected the actions and emotions of the infected, especially when we know that these mushrooms produce a chemical that goes into the production of the most powerful hallucinogenic drugs. that we know today.

There is another theory, according to which the dance group was a reaction to some tragic circumstances associated with this age, cases, wars, epidemics, that is, that anxiety and intense psychological outburst ultimately result in episodes of dance hysteria.

Whatever the reason, this dance epidemic was one of the strangest incidents in history.

Tantrum from laughter
In 1962, a girls' high school in modern-day Tanganyika, Tanzania, suffered one of the strangest epidemics the world has ever known.

The story began when one of the students told a joke to her classmates and they laughed at her, something normal that happens in all schools, but what is never unusual is that laughing at this joke lasts for hours, days and weeks ..

What a powerful joke! ...

This constant laughter made the parents of the students very confused, so why don't their daughters stop laughing ?! .. In less than a week, 95 out of 158 schoolgirls laughed nonstop, which forced the school administration to temporarily close.

But the matter did not stop there, as the infection spread to other schools, other cities and even other countries, and the number of infected girls and boys reached more than a thousand, forcing the government to close 14 schools.

The strange epidemic was marked by prolonged laughter for extended periods ranging from several hours to 16 days, sometimes accompanied by crying and intense emotions such as running hysterically, and most of those infected were teenagers aged 12-18. The epidemic lasted for about eight months before disappearing.

According to some researchers who have studied this epidemic, the reason for it is that students in Tanganyika felt very anxious, since in 1962 they saw the independence of the territory from Britain, and parents and teachers had high hopes that their children and students would to learn at the level of this event, which put these students under serious psychological pressure.

Halifax serial killer tantrum
Perhaps the most common type of epidemic hysteria among humans is fear hysteria. I remember when I was a kid in Baghdad there was a terrible serial killer named Abu Tabar who terrified every house in the city, and although I was very young at the time, I still remember that anticipation and horror on their faces when darkness fell. and how the streets were devoid of passers-by and plunged into darkness and darkness.

The city of Halifax in England experienced a similar horror in 1938 when two women said a man with a rubber mallet attacked them in the street. Quickly faced with assault charges by a mysterious assailant. Some victims said the attacker had a knife or razor.

The news of the unknown assailant spread like wildfire throughout the city and caused such horror that people even organized themselves into groups to protect themselves, and several strangers were severely beaten because people suspected them. The shops and shops were closed and the roads were cleared of passers-by.

Life in Halifax had stopped altogether or almost completely. City police struggled to find any clue that could lead her to her attacker, which eventually forced her to seek help from Scotland Yard.

Scotland Yard officers interrogated the attacked victims, and soon one of them confessed that he had not been attacked and that he made it up to attract attention, and confessions were made by others, ultimately showing that the attacker, in fact, it was not that all the stories were invented in one form or another, and that it was all nothing more than mass hysteria.
This incident is similar to another in London in 1788, when rumors spread of a large man attacking women with a small knife strapped to his knee and using it to stab them in the buttocks. More than fifty women are accused of assault by this man, and you are mostly beautiful women.

These attacks caused a wave of terror in the city, until women became afraid to walk the streets, and the very fact that any stranger approached a woman to speak to her became the cause of complete horror and could make the woman pass out from the intensity of her fear.

Newspapers of the time helped spread the story of this fearsome assailant after she called him "the monster of London." But in reality, it was just a paper monster, as police investigations revealed that most of the assault allegations are fictitious, and several female victims admitted that they cut themselves or cut their clothes to get attention, as it is the alleged monster. did not pursue or attack only beautiful ladies, as many women in London wish! ...

Although a man named Reinke Williams was arrested in 1790 and formally charged with some of these attacks and sentenced to six years in prison, most scholars today say the story of London's "monster" is wholly or largely fictional.

They already see .. Deja vu

You have never been in such a situation before, and then it seems to you that you have already seen this situation before, and you may be surprised by this strange feeling and feel that something is wrong! ... Don't be surprised, my friend, that feeling is a scientific phenomenon called the illusion of vision or deja vu .

Dejavo is a French word meaning "it has been seen before" and was first introduced by the French scientist Émile Bouarac in his book The Future of Psychology.

Those who have experienced this feeling describe it as the overwhelming feeling of being familiar with something that was not familiar to them before, for example, you can go to the country for the first time and take a walk down the street and suddenly feel like you've seen this place before, even though it your first visit. Or maybe you are having dinner with friends and discussing a certain topic, and it seems to you that you have already gone through this situation before: the same friends, the same dinner, the same topic! ..

Some scientists predict that 70% of people experience this phenomenon at least once in their life, and the likelihood that it will happen is higher for people aged 15-25 than for other age groups.

*In recent years, many scientists have begun to study this phenomenon, and with an increase in the number of scientists about the phenomenon, many theories have appeared, some phenomena are explained by malfunctions in the human brain, others are psychological reasons, and others say that the occurrence of the phenomenon is associated with the side effects of certain drugs, moreover. And we will look at the most famous theories in some detail. *:*

1.   The theory of one eye sees faster than the other eye
Proponents of this theory say that one eye sees landmarks faster than the other eye by a slight margin, that is, when you are somewhere, one of your eyes sends the coordinates and details of this place to the brain before the other eye, say, you entered street in Cairo, this street you have never visited and never seen in your life, but what is it? - The left eye will send the same details to the brain, and of course the brain will find that this information is already there and stored, and this will give the soul a false sense of familiarity with the place and the feeling that it was there before.

Against this theory, it says that she does not remember how she entered other senses in the case of Fish, and touch, you can feel that someone touched your hand before, or hear a song and remember that you heard it before. or you can even smell the perfume, which you smell for sure. Skeptics of this theory also said that De GAVO happened to some people who only see with one-eyed. In addition, this phenomenon does not always happen, that is, this feeling of familiarity does not strike you in every new place you visit, but it does happen sporadically from time to time.

2.   theory of psychological and mental disorders
This theory is outdated; doctors previously believed that the phenomenon was associated with mental illnesses such as schizophrenia and personality disorder, as well as in people with mild epileptic seizures. Scientists also say that the cause of the de GAVO phenomenon is a defect in the electrical discharge in the brain, many people feel a tremor moments before sleep, and this tremor most likely occurs during de GAVO, causing a false sense of memory. Those who feel these tremors all the time suffer from a recurrence of the de GAVO state, but it is mainly my voice, that is, he feels that he has heard this voice before.

There is also an opinion that this phenomenon is caused by the influence of certain drugs on the functioning of the brain, scientists say that
drugs amantadine and phenylpropanolamine for treating flu symptoms are the cause of the dgavo phenomenon , since dopamine affects some cells of the temporal lobe of the brain, and this explanation appeared in 2002. when two scientists observed a man who was taking the aforementioned drugs and showing dgavo symptoms .

3 - Theory of parallel universes

Parallel universes are a strange theory in every sense, which is difficult for the mind to believe, but the problem is that modern mathematical equations and physics prove their existence - at least theoretically - According to this theory, there is not one universe, but an infinite number of universes, and each of us lives with a different personality in each of these universes separately, and, oddly enough, all these universes can be combined and intertwined in the same place, for example, in your bedroom or on the street where you walk, but you you can only feel the facts and realities of your life in the universe ..

This theory is hard to digest, isn't it? ..

In general, in order not to delve into the lexicon of physics and mathematics, we will restrict ourselves to what interests us from this theory, that is, its relation to the subject of our research. According to Professor Steve Weinberg , Nobel Prize laureate in physics, there are several facts that exist together in parallel universes in the same place, and sometimes these universes overlap each other for one reason or another, allowing us to see facts that do not belong to us, and we have no previous knowledge about them, but belong to a different version of us living inside .. This explains the origin of déjà vu .

4-memory theory (information we knew and forgot over time)

Scientists say that De GAVO is the result of information that we learned a long time ago, but forgot for one reason or another, and now we have been able to restore it.

In addition, according to scientists, sometimes things can get confused in our brain because you may visit a place for the first time, and it happens that this place is similar in design, shape and color to other places that you have visited before, and your brain memorized its details and coordinates, and this leads to the imaginary feeling that you seem to see someone with a profile similar to that of one of your friends, and the sight of this person inevitably reminds you of a friend you know or even can imagine imagine .

5. Theory of Reincarnation

In some religions, such as Buddhism and Hinduism, and in the ancient Middle Eastern pagan religions, and in some sects, there is a strong belief in the reincarnation of Souls, which is that after their death people are born again in New bodies, that is, their souls are transferred from from one body to another, and this cycle continues only endlessly, or until the spirit reaches the stage of transcendence with its actions, its asceticism and its righteousness, in order to free itself from the cycle of reincarnation.

Simply put, reincarnation means that after death a person ends only his body, but his soul does not die, but passes into another body and is born again, in the form of a new family, and it is possible that a man is born as a woman in his new life. or a woman is born a man, or she will be a bad and broken family, or even be born a pig, or a dog, or a frog! ..

It is noteworthy that the mainstream and clear texts in the scriptures of the three heavenly religions, Judaism, Christianity and Islam, disagree and completely reject the idea of reincarnation and believe that a person lives only one life and bears responsibility and punishment for his actions only during this life. There are also many philosophical and scientific criticisms of the idea of reincarnation that cannot be raised in such a rush.

6. Theory of Dreams.

According to this theory, the world of dreams can sometimes be confused with reality, for example, you saw yourself in a dream walking along a certain street, and when you woke up, you forgot about the dream, but after a while you stumbled upon a street similar in shape, light or color to the street you dreamed about, and since dreams are remembered, of course, most people sometimes feel that they have seen a place, incident, or situation in their dreams before.

Finally, what do you think, dear reader, what theories do you like, and have you ever had the de GAVO phenomenon? !!

Memory

We all have a huge store of memories that have accumulated - and continue to accumulate - throughout our lives, ranging from happiness and beauty to sadness and depression, and we continue to remember some of them, while many - if not most - languish in subconscious, hidden from view.

And we all need memories, We see no doubt that what is happening in its entirety is free, we argue about how realistic this madness is, but scientific studies say that some of our memories, which we take to health in full, may contain some scenes and events that did not actually hit, they may differ from nowhere .. I see the surprise on your faces.

This is due to internal or external suggestibility, which causes what is known as "false memory", the term first appeared in the words of the American researcher in human memory and cognitive psychology "Elizabeth Loftus ", and "false memory" can define as memories that were distorted or "fabricated", and in some cases the memories were originally invented, and this term differs from the term "false memory syndrome", since the syndrome is more influential and more severe in the life of its owner, of course.

At the end of the last century, the aforementioned researcher conducted a study to test the degree of influence of suggestibility on the reliability of memory, and she showed participants a video showing the effect of self-driving cars on each other, then divided the participants into two groups and asked the first group: "how fast were these two machine! "while another group was asked the same question, saying" I broke some "replaced with" I hit a few. "

A week after this study, the participants were recalled and the same question was repeated to them, in addition to another question: "Did you see a broken glass in this video ?!" The group that was asked the first question "broke a little" was more likely to answer "yes" to the second question, as the wording of the question suggested the severity of the encounter, as opposed to the other group, which tended to be negative.

Several years ago, the same researcher, in collaboration with the online magazine Slate, conducted a large-scale study of over 5,000 people to test their ability to remember certain political events in the past.

The study is based on showing four photographs in front of each participant, one of which is fabricated, and asking them if they remember what is shown, and the fabricated photograph was shown by former US President Barack Obama shaking hands with former Iranian President Ahmadinejad .

Surprisingly, more than half of the participants in this study responded that they remember what appeared in the fabricated painting, and some even described the event by talking about their feelings at that moment.
Through the aforementioned research, Elizabeth has demonstrated that memory can be deceived and deceived by suggestion, be it puns or scenes, as has happened in the aforementioned research or others.

In a remarkable study of 1,600 students from the same university, researchers from the American University of Gola found that 20% of students had memories that had no basis in reality, that is, just an illusion.

It should be noted that the theory of "false memory" has been subjected to a campaign of skepticism on the part of some lawyers, since acceptance of this theory, in their opinion, will diminish the value of testimony as forensic evidence.

The parasite that controls your brain.

It's time to enjoy some monster stories, the scariest ones are the monsters that already exist.

Join us to share some of the scariest stories of parasites - those who control the mind of their human host, sometimes leaving madness and death behind them, these creatures that live and control the bodies of beings bigger than them.

Sly parasite
Toxoplasma gondii Toxoplasma is at the top of the list of the most famous - and most controversial - parasites.

This tiny initial doesn't look much more than a bubble, but once it hits the brain, it can radically change the behavior of its host mice, cats, and even humans.

The original hosts of Toxoplasma are felines, where their eggs (known as "oocytes" or "eggs") are deposited in the feces of the cat, waiting to be picked up by a vector such as mice, once they are safe and warm in the intestines of their temporary host, while the eggs won't multiply quickly, those small, modest dots that can actually do some damage migrate into the intestines.

But when the moment is right to strike, the tiny Toxoplasma changes its host's brain chemistry, the infected mice lose fear of their feline nemesis and even become attractive to its scent! She fearlessly leaps into his claws, where she dies and returns the Toxoplasma to the cat to begin the egg-laying cycle again.

Quite scary, but not to the same degree nightmares, mice are not the only hosts that Toxoplasma can live in, and some researchers estimate that up to 30 percent of the people on earth - more than two billion of us - will carry Toxoplasma in their brains.
And what happens to people with toxoplasma? They may become less risk averse, their reactions may diminish, or they may become very sensitive ... In severe cases, this can cause schizophrenia.

Some studies have shown that the incidence of schizophrenia increased dramatically at the turn of the 20th century, when domestic cat breeding became commonplace.

Another article, published in the journal Royal Society B, argued that in areas where Toxoplasma infection is high, these small parasites can cumulatively alter the behavioral patterns of entire cultures, and researchers found that infected parents have a 30 percent chance of transmitting the parasite to their children. ...

And look at this: Researchers estimate that over 60 million people in the United States alone are currently carrying Toxoplasma, and most of them have no idea because these parasites often cause no symptoms at all .. Until the day when you hit.

Madness of the amoeba
If you're out in the wild, stay away from fresh, warm, stagnant bodies of water, no matter how thirsty, these little ponds are often home to Naegleria fowleri chicken Naegleria , a type of amoeba that loves human brain tissue, spends long periods resting as a bag that looks like a small armored ball that resists cold, heat and drought.

As soon as they enter the host's body, until they move and transform into a form called atrova (the active role of the first life cycle of an animal), and as soon as they transform, they go directly to the host's central nervous system, obeying nerve fibers in search of the brain , and once they settle, until they start dividing, dividing and multiplying.
Symptoms begin with changes in taste and smell, possibly some fever and stiffness, and over the next few days, constantly digging into the brain, the victim begins to feel confused, has problems with attention, begins to hallucinate.

With the loss of control of the brain, unconsciousness ensues, and the victim is only two weeks away from death, although one person in Taiwan managed to resist this for 25 days before his nervous system stopped.

Although anger injuries are rare - there are only hundreds of them in world history - they will always be fatal and difficult to heal before they get out of control, but it would be wise to avoid warm pools of standing water to avoid being an intruder in your the brain.

The virus that causes fear
We have always been forbidden from approaching wild cats and dogs, and also not to disturb the animals that roam the city streets, they can seem friendly, they can carry the deadly rabies virus, which does not always cause foam at the mouth, but alters the brain functions of its victims.

This bullet-like virus is very small and skillful, it always manages to break out of the body's immune system, and it does not need an invitation to enter the body of a new host, a simple wound is enough, and once it reaches the bloodstream, it will attack the cells and turn them into factories. for the production of rabies, which produce thousands of copies of the virus, and as soon as the attackers grow in quantity, until they make their way into the host's nervous system and into the brain.

But rabies viruses don't settle anywhere in the brain; they specifically seek out the hippocampus , amygdala and thalamus, brain structures that play a significant role in memory, fear and emotion.
They don't randomly engulf brain cells, instead they alter the way cells release neurotransmitters like serotonin, GABA, and endogenous drugs - in other words, directing their host's brain chemistry against it.

For animals infected with rabies, they most often attack all living things near them, but this is more fear than anger, and for people infected with rabies, they will be afraid of water and air gusts, which makes them retreat and shiver, unable to control themselves.

If infected people are not treated quickly, their bewilderment and confusion will intensify in addition to hallucinations, violently attack the threats they perceive and unhappy passers-by, they will lose the ability to sleep, sweat profusely and ultimately become paralyzed as their brain functions turn into chaos. and after a few days the paralysis will reach the heart and lungs.

So far, less than 10 people have survived a dangerous infection with rabies, these ten cases are the only historically recorded worldwide, and many doctors consider the disease incurable, and it is good news that although the disease can be easily prevented with a vaccine if you plan go to any place where wild animals roam, it is best to be prepared.

Sleep parasite
In sub-Saharan African villages and the wilds of the Amazon, a very small insect can bring sleep that leads to death, the tsetse fly loves the taste of human blood and often carries a parasite known as Trypanosome or Trypanosome, which tastes like the human brain.

The life of parasites of the trypanosome genus begins in the intestines of invertebrates, but develops quickly when they come into contact with mammalian fluids, at the initial stage of infection, the parasites live in the blood and lymph nodes of the host, where they grow from small ovals to something resembling a stain, equipped with flagella ...

As they grow older, parasites cross the blood-brain barrier, change the structure and function of their host's brain cells (parasites seem to have a special attachment to the hypothalamus, which helps regulate our mood and sleep-wake cycles), and the host begins to feel and behave strangely, first in he has headaches and trouble sleeping, or sleeps and wakes up at night.

The host person is exposed to an amazing variety of other psychological symptoms, from changes in appetite to depression, strange speech patterns to uncontrollable itching and tremors, and over the past few years, the host's strange behavior has begun to gradually shrink to laziness and unresponsiveness, and, finally, prolonged sleep leads to coma and death, hence the name "sleep disease".

While there is treatment for trypanosomiasis, friends and family of victims often fail to catch the disease early for a very simple reason: the patient's wide range of variable symptoms make it difficult to recognize the disease.

If you have a friend who suddenly starts waking up at odd hours and eating less food, your first thought is, "Maybe he has parasites invaded his brain?" .. - Of course not. You will definitely think, "Maybe he's depressed," and this is what the friends and families of most trypanosomal hosts thought ... Until it's too late.

This is indeed something weirder than fiction when it comes to these parasites that take over brain cells and escape insanity.

Cheryl .. And the heart that loved her twice !!

- Help me .. Baby, please .. I'm dying.

These were his last words, which he uttered after he decided to end the life of people who had never been the same .. But who is he?　ˤ.. This is Terry Cottle , the man who silenced his heart twice .. How is it? Come with me to learn the story of a heart that was deceived by the same woman .. Twice !!
Terry Cottle lived a quiet and stable life with his wife and
two daughters in Jasper County in the US state of Georgia, until one day his wife received a call and was told that her husband was having an affair with another woman. And this woman's name is Cheryl .

Cheryl divorced her husband shortly before she accused him of having an affair with another woman, but it turned out that she was the one cheating on him with Cottle , prompting the deceived husband to call Cottle's wife to tell her the truth .. This contact put an end to Cottle's infidelity to his wife. and they got divorced ..

In May 1989, Cottle's relationship with Cheryl went from private to public, where they officially married and together began a life full of hope and ambition. Cheryl studied to be a nurse while Cottle worked, and they had a daughter, Jessica, who lived with two children. Cheryl from his first marriage, whom Cottle considered his sons ..

In late 1994, Cottle was able to move with his family from one small house to another, he developed himself despite graduating from high school and was able to obtain a license to work in real estate, then became an emergency medical technician, and Cheryl became a nurse. But despite all this, their happiness did not last long, and problems began to appear, after which Cottle left home and moved to live with his sister, but after a while they made up, and Cottle returned home .. However, this reconciliation lasted no more than three weeks with Ever since it happened on March 15th, a furious altercation between them dumped Cheryl during her wedding ring and threw it away, saying that she couldn't live with someone who makes more money from it !!
The next morning, they agreed to end it all after their relationship had reached an impasse, and when Terry was about to leave, he went to the bathroom and shot himself with a pistol, believing that by doing so he committed suicide and silenced his heart forever. said that the bullet entered his skull behind his right ear, but did not come out on the other side and as a result entered the stage of clinical death, and after 4 days in the hospital, Cheryl agreed to separate the devices from him and also donate his organs,

despite the objections of her father .. Thus, Cottle's heart is destined to live another life in the body of another 57-year-old man named Sonny Graham.
Sonny Graham lived in Hilton Head , Charleston, a man known in the area for many years as the manager of a telecommunications company that provides telephone lines to all residents, he was a life-loving person as he had a special presence at all social and charitable events. he loved golf, he was a key player in the local team, in addition to his love. In his personal life, he was an exemplary family leader, as he and his wife Eileen were a wonderful couple for over three decades, resulting in two sons, Michelle and Gray.

But in 1994, Sonny Graham contracted a virus that damaged his heart, and the doctor told him that he had no hope of life other than a transplant, and so Graham was placed on the list of people in need of a heart.

Graham delivered his command to God and drew to a close, until, after a year of waiting, he received a call that served as a reminder to bring him back to life .. He knew he had a heart, and of course it was not that other than the heart of Terry Cottle , who died of suicide by shooting himself in the head.
The heart transplant was successful, and a few DAYS Sonny Graham was reborn with a young heart, who was only 33 years old .. And six months later he was able to return to what he was before, and already went fishing with some of his colleagues .. And his friends noticed some changes, such as his habit of drinking beer again and his insatiable eating sausages, which became his favorite dish .. In short, Graham seemed like a man who wants to come back to life again !!

In November 1996, Graham decided to send a letter of thanks to the donor's family through the South Carolina organ donation agency .. He didn't know that his decision opened the door for Cheryl - the man hunter - to enter his life and break his stability !!

After exchanging several messages, they agreed on a meeting, which took place in January 1997 in the presence of Graham's wife Ellen .. Throughout the hearing, Graham could not take his eyes off the young widow .. Later he said that he fell in love with her at first sight and even felt that that he has known her for a long time ..

This encounter was followed by a series of encounters that knocked Sonny out of his mind and he no longer only listens to the voice of his heart .. His heart, which was actually connected to this woman before, when he lived between the ribs of his original owner Terry Cottle !!

In April of that year, Cheryl , then 30 years old, married a third time to a man named George Watkins , and both Graham and his wife were present at the wedding due to at least an innocent friendship on the part of his deceived wife Ellen. .. But in 1999, after Cheryl gave birth to Watkins , Ellen realized that her husband's relationship with this woman was never innocent, and so in 2001 Graham and his wife were officially divorced, which revealed a 38-year relationship, and the family, which everyone thought was ideal, fell apart.
Right after the divorce, Cheryl moved in with Graham .. But can you continue with him without problems? In 2002, Graham sued her, accusing her of giving up some loans and

not returning the diamond ring he gave her, accusing him of threatening her when she asked him to end their relationship. ...

Cheryl got a formal divorce from her husband Watkins during her disagreement with Graham , and not only that, she married another man named John Johnson, who was a guard at the Georgia jail where she worked as a nurse .. I don't know, as if this woman was married and then divorced .. A year after she and Johnson got married, problems began to arise between them and she accused him of domestic violence, and they eventually divorced in 2004. During Johnson's divorce proceedings noticed that Cheryl was wearing a diamond ring that Graham had given her , indicating that the water between them had returned .. Already on December 8, 2004, Graham married Cheryl , and they moved into a house designed the way Cheryl wanted , and Graham accepted the job of her two sons at a design company he just founded ..

A few days before their second wedding anniversary, Sonny and Cheryl attended a ceremony honoring organ donor families and they were in full view of the public and wrote an article about the unprecedented love story and destiny that brought Cheryl together as someone with the heart of her ex-husband.

But despite Sonny Graham's seeming joy , he was found dead a few weeks later in the backyard of his house after shooting himself on an overcast and gloomy spring morning with the gun he always used on the hunt, and, oddly enough, the bullet hit him on the right side of the neck just like it happened 13 years ago with Terry Cottle !! And shut up my heart again, but this time forever ..

The news of Graham's suicide shocked everyone who knew him, a man who always loved life and a few days before his death went fishing with friends, where he talked about his future plans, but they did not know that in the last days of his life he wrote a will and made Larry his nephew.
Although the coroner determined in May that the death was due to suicide, the investigation remained open and Cheryl's surviving spouses were summoned for questioning. Johnson mentioned a fourth husband, that there was an incident that caused him to hasten to break up with her, namely that one night she talked about suicide and she went up to the bathroom, and when Johnson followed her and her grandfather, threatening to shoot himself with a .22 pistol, he came over to take the pistol from her and there was a fight!

Her first husband said that in 2005, she threatened to blow off his head with a pistol after a fight over custody of her grandchildren!

The question of the pistol and its reference has raised many questions more than once, especially since Terry Cottle 's suicide contradicted her testimony, where she said during the initial investigation that she heard her young son screaming that Cottle had shot himself and when she went to the bathroom, she found him wounded and lying on the ground with a pistol in her hand, but she changed her mind.

It is worth noting that Cheryl's actions after Graham's suicide were not those of a wife grieving the loss of her husband, and may have been mostly married to him because of his money, she complained to one of her friends that her husband did not leave her a penny. and the fact that Graham died and left behind a lot of debt, which surprised his nephew Larry.

Cheryl was indeed a strange and confusing woman, her parents said that she posted a photo of the man on her account on one of the Sites a few days before Graham's suicide , that he was her new lover !!

Graham's suicide left behind many questions related to the theory of cellular memory, which claims that human memory, with its habits, qualities, feelings and tastes, is connected not only with the brain, but also with cells of other organs of the body, such as the heart. what happened to Graham was because of Cottle's heart implanted in his ribs, or what makes him fall in love with the same woman Cottle has already fallen in love with and married, and then commit suicide in the same way ?!
Is the heart just a pump for supplying blood to the human body, or does it really retain some of the memory it transfers to someone else's body when it is transferred and transplanted? ʕ

In fact, there is a heated debate between proponents of the theory of cellular memory and its supporters. In this regard, Dr. Paul Pearsall conducted a study of many cases of heart transplants, and in 2002 he wrote a book in which he talked about the cardiac code and gave several examples, including a 47-year-old woman who had the heart of a 23-year-old man who died as a result of a bullet in his lower back .. This woman says that after the transplant she had persistent back pain !! And a 52 year old man who had the heart of a 17 year old boy who listened to loud music after loving the silence and also started acting like a teenager !!

But skeptics of this theory say that the heart receives its commands from the brain, that is, the brain is controlled by the heart through the nerves, and that the tension and increase in the number of impulses in the heart is only the result of what the brain experiences from emotions or pressure, and all research on the heart does not could do it. Scientifically, the heart's primary function is to pump blood and distribute it throughout the body.

But far from science, which only recognizes physical evidence, we find that the heart has been held responsible for feelings and sensations for centuries, and often we turn our backs on things just because our hearts are unsure, and to this day there is no scientific evidence to prove it. that the mind is focused on the brain .. And if we turn to the Holy Qur'an, we find that Almighty God spoke to the human heart in more than one verse when it is used in Surah Al-Hajj: "walk the earth so that hearts come or ears heard that they do not blind the eyes, but blind the hearts that are in the chest. " "

"But believers who, if God is mentioned and their hearts are pure, and if his verses are read to them, increase their faith and trust their Lord."

We see through sacred verses that Hearts reason, are in confidence and foresight, therefore their function is not limited only to providing the body with blood ..

*And now back to the heroes of our article .. Did Cheryl deserve the fact
that Cottle and Graham - both unfortunate men - left their lives, stability, wife and children for her? In my opinion, she is an extremely exploitative woman, they may have fallen in love with her, but I don't think she reciprocated them, they were just numbers in her life.*

This begs the question: Was Graham's suicide actually caused by Cottle's suicidal heart memory , or was it related to Cheryl and her psychological revelation she may have experienced on her husbands when she was bored and set out to get rid of them ?! Or is there a third reason that no one knows about .. What do you think, dear reader?

Metomania .. Mania of lies

To begin with, we must all admit that the lie has influenced us in one way or another, to one degree or another between us, and that no one has experienced a lie in the course of human history, but the question is the same here .. Why are We lying ?!

We usually use lies as a lifeline or bridge to get to where we want to go; we can use it to avoid crises that can escalate into sinister ones, or to maintain certain social relationships. we can use it to achieve certain personal interests or material benefits. in this case, it becomes a justified means of "Machiavellianism" to achieve goals that may or may not be noble. in any case, the above and similar motives may - at least - be reasonable, although the lie is rejected.
But what if the motivation to lie is free from coercion, greed, or any of the conventional reasons ?! What if one of them continues to commit a lie, the avoidance of which cannot cause the slightest problem to him or others, or pose any danger to his interests, status or relationships ?!

Here, a lie can turn into a pathological condition known as " metomania ", "pathological lies" or "obsessive lies." This condition was first described in 1891 in medical literature by the German psychiatrist Anton Delbrück .

I regret to say to those who are bored of reading that the past was nothing more than an introduction, and I am happy to invite those who like to read to finish it to dig deeper into our topic, let's continue ..

Metomania
By studying many of the writings on metomania and connecting with those who, in my opinion, are at least one of its owners, I can define it as a chronic psychological error that prompts its owner to come up with stories or events that often revolve around it, which can be exaggerated - sometimes strongly - to satisfy purely psychological desires, and these fabricated stories may be based on real facts.
The author of this disease may be aware that he is lying, he may think that what he is saying is the truth itself, or it may be that the "truth" is just an old lie that repetition has made a reality in his eyes through self-suggestion, and so he fell into the trap of "false memory", I touched on this topic in the previous article.

I think that many of you have gathered to meet with one of the victims of the "pathological lie" - at least - and you may have noticed that his speech is usually focused on himself, he can talk about his heroism, or about his courage and progress, or about his relationships with the owners of matter and person and glory, which arose only in his wild imagination!

As I said before .. Talking about a "pathological liar" can include some exaggeration that makes this conversation funny and ridiculed by others, which he vehemently rejects, and even fabricates lies just to reflect the character that should not ridicule or underestimate!

Lying geek .. Why is he lying ?!
The reason is that lying gives him a certain psychological euphoria, he does not lie, greedy for money or Jah or position, he lies because he deceptively believes that he finds himself in a lie until the lie becomes an existential matter for him. he lies to appeal to the emotions of others towards him; he lies until he gets their attention and gets their support.

Try to explain the psychological geek with books, says psychologist Dr. Claude Bilan: "Dastardly people don't want to disappoint others, prompting them to exaggerate their statements, even react to what others expect."

One study found that nearly one in 1,000 young offenders have the condition, that the age at which the disease begins is 16, and also found that people with the condition have average or above average IQs and that they have high verbal skills, excellent from performing skills.

The study also found that 30% of people with the condition grew up in a chaotic home environment where a parent or family member has a mental disorder and 40% have a nervous system disorder that is epilepsy, ADHD, or others.

The signs that appear on the liar
When a liar begins to invent his lies, vital changes occur in his body that quickly impose signs on the behavior of the owner, so an expert can reveal his lies without much effort, and this is true for all liars in general, including those who are obsessed with lies.

One of these vital changes, the body releases adrenaline, which, as soon as it enters the capillaries of the nose, until the owner begins to touch his nose, and the substance adrenaline irritates the salivary glands and then often dries them out, which leads to accidental ingestion on an empty stomach.

Also a sign of a liar is that he puts his hand on or next to his mouth - for example, on his chin - during a conversation .. As if he is fighting with his hand, and she tells him to lie and tries to silence him!

When someone wants to recall real events, then this is - in general - a system from above to the left side, the head of one system from above to the right, it is possible that a creamy lilo is preparing to lie on the state, except for the left-hander - I am one of them - the above will be upside down for him.

Below is a short list of other signs of a liar:

- Avoiding eye contact with the listener.

- Development is proceeding at a faster pace than usual.

- Fidget.

- Stuttering or stuttering.

- Frequent terms.

- Sweating.

Important note: Some people have some signs of lying does not necessarily mean they are lying, and rushing to sentencing usually leads to serious consequences.

Pathological lies and obsessive lies
Typically, the terms "pathological lying" and " compulsive lying" are used in medical circles for the same purpose and interchangeably, while some experts believe that " compulsive lying" differs from "pathological lying" in several ways, including:

- That a compulsive liar cannot control his tendency to lie, and he feels more comfortable when the lie affects the truth.

- That the obsessive liar lies on matters both important and unimportant.

- That the obsessive liar has no ulterior motive behind his lies.

- That a compulsive liar can hardly tell the truth, even after being exposed.

This is the other side. Professionals may have difficulty diagnosing metomania due to its disappearance behind many mental disorders as a symptom among many common symptoms, so they take training courses that enable them to recognize and diagnose this disease, and the American Society of Psychiatrists included " metomania " as a mental disorder in the third edition of its manual, which is the world's first reference to psychiatry, called " metomania ".

Light speed: court time

After the Industrial Revolution, man was able to overcome many of the extreme speeds of the creatures around him, he overcame the speed of the Falcon up to 300 kilometers per hour, the fastest living organism on our planet, and he was able to overcome the speed of sound in air, which is 1224 kilometers per hour, and the Leap Felix is just Felix's jump.

In the midst of breaking these records, he faced a huge barrier that cannot be met and surpassed the maximum speed in nature.And even the maximum speed reached by man is a trivial thing in front of this barrier, which is called the speed of light!

Light speed:
Initially, we can define light as the energy of electromagnetic radiation, which includes those that fall within the range of the human ability to see, and light has a speed that is slightly dependent on the presence of other elements, such as air and glass, but is fixed in a vacuum.

In the second half of the last century, after many very accurate calculations, scientists estimated the speed of light in a vacuum at about 299,792,458 meters per second (about 300,000 kilometers), which allows light to circulate around the planet three times in a row — in about just one second.

Now, dear reader, you can say: what if I could travel so fast ?!

Keep reading the article and you will be surprised at the answer ..

Light year :
Due to the enormous distances between the elements of our vast universe of planets, stars, galaxies and others, scientists have adopted the so-called "light year" as a unit of measurement for these distances, and a light year is the distance that light travels in a vacuum throughout the year ..

Ok .. If you don't have a calculator that can display astronomical numbers, don't bother with calculations and even save yourself the hassle, the speed of light in a vacuum for a whole year is - I'll have to write this number on a new line ..

800 580 472 730 460 9 meters, that is, more than 9 trillion kilometers!

By analogy with a light year, distances can be determined in minutes, hours, daylight hours, etc.

The sun - for example - is about 8 minutes from us, that is, the distance traveled by light in 8 minutes, and if we assume that someone is calling you from the sun - unless it evaporates on its outskirts, then if it starts speak: Hello, you will not be able to hear it before 8 minutes have passed, and in turn your answer will be heard.
Until the picture becomes clear, when you turn towards the sky to see the sun, you do not really see it as it is instantaneous, because the state of the sun that you see at this moment is eight minutes behind its reality, and so the same is the case with stars - for example, if we assume that there is a star about 50 light years from Earth, the Earth and we were able to see it, and to find out how it looks at the moment when we see it, we will have to wait more fifty years!

Have you ever thought about how long it takes for a response to come from you and from you with someone who calls you from this Star ?!

I don't think we have enough to complete the challenge!

What if we could travel at the speed of light ?!
According to Einstein's theory of relativity, there is an inverse relationship between speed and time: as speed increases, time slows down and contracts until time completely stops at the speed of light - in a vacuum - and the question that might come to mind is, what if we can travel with such speed ?!

Suppose you have a vehicle that travels at the speed of light and you decide to travel to a star 1436 light years distant from Earth, and when you got to that star, you used a telescope to see everything that happens on earth. do you know what you'll see ?!

For example, when you go to the Medina area, you can see the Prophet (peace be upon him) and his companions preparing for the battle of Badr with the Quraysh , or you can witness the joy of our Prophet by turning the Qiblah into the Kaaba .. In other words, you are watching the Earth in the second year of the resettlement, or in 623-624 AD, and wherever you roam with your super-telescope around the planet, you will not see anything of what is happening in the modern earthly world, but you will see the state of the Earth in 1436 years !
Well, let's say you quickly missed returning to Earth, a planet that includes your people, your loved ones and everything that dies for you, so you decided to leave this lonely, dormant Star and return to the planet you belong to .. I'm not exaggerating if you say you'll be shocked by what you'll see when you arrive ..

In fact, several centuries would have passed since I left the planet Earth ... And you had very little time, maybe a few days!

The fact is that while traveling back and forth, time was suspended for you due to the speed of light at which your ship flew, but not for the inhabitants of the planet, and because of the short period that you spent on this star, your age increased slightly as the planet passed centuries equal to the distance of light between it and the star.

What if you want to use the speed of light to travel the planet ?!
Let's say you went with your parents to one of the stations of the high-speed train, and then they said goodbye to you at this station, wishing you a safe arrival, and did not take their eyes off you until you got on the train, which quickly moved off .. So now will be?!

What happens is that you see your parents leave the station even though the train is moving, and the reason is that its speed is comparable to the speed of light, which makes the train look like it did not leave at all, so time stops. for you and you can watch your parents leave the station!

Sorry, however, for the digression .. What if this train moves faster than the speed of Light, will time appear again ?!

I will not be surprised if I answer that you will not only be able to see your parents, but also see yourself saying goodbye to them at the station .. In other words, Time will begin to return to the past!

Wheaton : the mystery of the lost planet

In 1975, a geology student in the Farfum Desert in northeastern Ashgabat, Russia, found a pile of glass of unique colors between green and black and the size of a coconut grain among the desert sands and quickly handed it over to the expedition leader (Pavel Florensky), who took the glass shards to Moscow. ... - Tektites, which are lumps and fragments of a dark bottle that supposedly came from space and includes known minerals, including beryllium, and tests showed that these minerals turned into glass after exposure to immense heat, these tektites reopened a long-forgotten issue of Viton's theory about the Lost Planet.

In order to know what the Lost Planet Theory is, we must first learn to code (Titus - bode) an alien. In 1766, the German mathematician Johann Tete Virus, then director of the Berlin Observatory, noted that the dimensions of the planets from the sun and the distances between the planets of the solar system obey the law of the athlete; we can speculate that we define after each planet from the Sun, before the discovery of the planets Uranus, Neptune and Pluto, he staged a series of sports as follows:

0 -3 - 6 - 12 - 24 - 48 -96 - 192 - 384

These numbers are multiples of 3.

Then the German astronomer Johann Bode appeared in 1778, he said, adding the number 4 to the row to read the following:

4 - 7 - 10 - 16 - 28 - 52 - 100 196 - 388

I raised this law in a moment of surprise at the iPod standards , insofar as it applies to the actual sizes of planets from the sun, which are measured in astronomical units and are as follows:

Mercury (4) - flower (7) - floor (10) - Mars (16) - Saturn (52). customer (100)

But, I remained three sites free and (28) and (196) and (388) this was a disregard of the law until the planet Uranus was discovered in 1781, and after the Sun is as much as 191.8 and there is an asymptotic number of 196 in the iPod law , which renewed interest in this law and the extent of your health, and increased attention when Pluto was discovered in 1930, and as much as after (in time) the sun at $ 394 and there is a number asymptotic to 388, aged attention this zakonu.eto law, but lost there a space, and is the number (28).

Fitton : the mystery of the lost planet
Cyrus : a small asteroid in the asteroid belt between Mars and Jupiter
In 1801, an Italian astronomer (Biase) discovered a small asteroid orbiting between Mars and Jupiter - and his name (Ceres Ceres) is also 27.7, and this figure is very close to the number 28, but he is an asteroid, not a planet, from for which it was theoretically considered emptiness.

Today, most astronomers accept the iPod's law and consider it just a coincidence, but this law looks like a miracle, and the laws are unusual, maybe just a happy coincidence, although most of its results come much closer to the real differences? God, I know, but *what interests us here is the mystery of the lost planet between the orbits of Mars and Jupiter .. what happened to him, and there was already a planet in this place? ..*

Fitton : the mystery of the lost planet
Planet X or Viton .. they say that it was somewhere between Mars and Jupiter
Some scientists say that yes, there was a planet in this place, this planet, the names of several planets X and Viton , Nibiru and Marduk, but what we have now is just a belt of asteroids scattered here and there, where did the planet go ?.

Scientists of the former USSR say in their assumptions that the lost planet was similar to the Earth in terms of climate, and was named an academic (Sergey Orlov) name (Viton), but that the planet completely faded from the explosion and was not followed only by the asteroid belt, the most a large bowl called Ceres.

Fitton : the mystery of the lost planet
Agglomeration jackets: stones glass source by space
An important part of the theory of the lost planet (sulfate agglomeration) rose higher and was found in various parts of Russia, Australia, the Philippines and Indonesia, as scientists interpret, the appearance of such stones is a really massive explosion caused by the heat of millions of sorting works to throw stones into space and on The earth.

There is other evidence for the existence of a lost planet, but for this evidence you don't have to go to the space research agency - NASA - or knock on astronomers' doors for your giant telescopes, you have to go to the Archaeological Museum to underestimate the Sumerian clay tablets !! ..

But what do the Sumerians, who appeared and lived on the banks of the Tigris and Euphrates rivers for more than seven thousand years, do with this table of the depth of scientific astronomy, which, as you know, is a relatively recent science? ..

To find the answer, we will have to plunge a little into the myths and literature of the Sumerians, where you wrote Sommer's poem on a clay slab a thousand years ago that a celestial body called Nibiru appeared in our solar system .. and how the poems tell us about the appearance of this planet that between Mars and Jupiter there was a planet called " Tiamat ", and that the planet suffered a skeletal catastrophe as a result of a collision with the planet or other major crime that came from outside the solar system. This collision severely destroyed Tiamat and blew it apart to be some kind of Earth, while others turned into a strip of small asteroids, orbiting during the day between Mars and Jupiter ..

Fitton : the mystery of the lost planet
Sumerian seal of the third millennium BC and traces of great noise, where you show - as some say - the Sun, planets - the solar system - and shows the planet X is Viton or a mysterious planet .. note that the Sumerians did not have telescopes like were they able to draw the solar system ?!
And tell us a fairy tale or legend that the other planet that I hit Tiamat was the largest of them many times the size of a palm and therefore you survived the impact catastrophe

and did not budge, but the force of the impact affected the orbits, so it did not was able to get rid of the gravity of our solar system and became a part of it, and its orbit is similar to the orbit of many famous notes and the one that about the sun takes, according to the Sumerians, about 3600 years.

An echo of this legend, the ancient Sumerian frequency later in Babylonian mythology, in particular in the myth of the creation of the world, the Babylonians became famous (Inoue as Il-debate), but the Babylonians called the planet Marduk, and not Nibiru , which made the legend confusing and a bit mysterious, because it is known that the Babylonians called the name of Murdoch on the planet Jupiter, was Tiamat colliding with the planet Jupiter? .. And the buyer, as you know, of the planet Titanic, the largest in our solar system, inevitably went from such a collision due to the destruction of the planet itself to another.

Fitton : the mystery of the lost planet
The Sumerians said that the gods descended from the sky and taught them civilization and wisdom
One way or another, despite the difference in names, the Sumerian, Babylonian inserts, both of which refer to access, collisions between the two planets led to the appearance of planets and other worlds, including the planet Nibiru , and the image of a mysterious planet clearly appears on some seals of the Sumerian image of the solar system and shows us the presence of ten planets .. which is very strange, because we are talking about people who lived thousands of years ago and did not have telescopes and astronomical observations and modern ones.

And this is where the story ends, it is true that the Sumerians disappeared a very long time ago, but this alone suggests that she is still a creamy substance of mystery seekers, which appeared in 1976, the writer controversially calls in America " Zechariah will be" and tries to benefit from of this ancient legend, tell them a new version of the story of the lost planet, also called Planet "X-X", after mixing information and leaves.

Today I have many followers of this writer who believe that his word is reality and not pure science fiction.

Fitton : the mystery of the lost planet
Many: they have landed from the sky.
Let's take a look at the background or story that came out of this man based on the legends of the panels and Sumer. Many Sumerian sources say that their ancestors received from the hands (goddesses) who came from heaven to earth, some departed from the Anunnaki and mean in Sumerian: (who landed from heaven to earth). And tell us the mythology that these alone they came from a planet called Nibiru , the planet we just talked about, landed these objects on earth before 445 thousand years in search of natural resources - like in the movie Avatar ! - Especially gold, which they found in large quantities in Africa. But over time, the hair of these extraterrestrials got bored with mining work that does not belong to them, so they decided at some point to create objects, imputed to them as slavery, and so they created humans through genetic engineering, creating their genes evolving with the genes of ancient man -specialist, like a monkey, who lived like animals in

caves in those distant times of the past, this led to the hybridization of the genes of modern humans of the modern type, which we humans exist ..

Fitton : the mystery of the lost planet
They helped people and their world with knowledge, wisdom and civilization
Stories that only they became gods for people and help them in promoting and teaching them the origins of civilization, and thus the early centers of human civilization appeared, and it reached the Sumerians at a high stage of scientific progress and cultural and artistic, where they opened the wheel and they write many things, the Sumerians tell us through clay tablets that they bear a lot of them alone. This also explains the great boom that happened to humanity five thousand years ago, when they suddenly entered an advanced stage of civilization, especially in the valley of my Mesopotamia and the Nile and, after hundreds of thousands of years of life, in the cave of darkness!.

Well if you could give the validity of this story or premise, where are they alone today? ..
Fitton : the mystery of the lost planet
Where are they now? ..
Zakaria says it will be that destructive wars arose between them because of greed, and because of the results of these wars, the destruction of the city of Ur, Sumerian, nuclear-2000 BC. e.!. Fatal wars have left Earthlings alone in the world they come from.

In an attempt to validate his theory, Zechariah is trying to establish a connection between Anna and the sex fleet in the world mentioned in the Old Testament and the race of giants caused by the increase of the Sons of God - angels with human women. You may know that the sons of God are exiled from the earth because of their relationship and their desire to make women human beings.

But did you end this story here?

No, there is more of Mr. Zakaria in his quiver , he says that Anne, who gave us part of her genes and put our feet on the first trail of civilization, will one day save humanity from the destruction caused by man himself. Here we go back to the story of the lost planet Nibiru , which dates back to the solar system every 3600 years, this planet will probably return, bringing with it a lot of bones to save us again. Here you see that Nibiru will be pregnant with a catastrophe for the Earth, possibly due to a collision, and that this is inevitable and will happen in the near future.

Fitton : the mystery of the lost planet
There are those who believe that Nibiru will return, bringing with it the end of our planet. In fact, the story of Zechariah will not be a beautiful story, but a science fiction book in which all the factors are in awe, that we love to read about them, about an old lost planet suitable for human life, but it was life, but it suddenly disappeared and its survivors the inhabitants migrated to planet Earth to be (the gods) that servant of the ancient people, and they sent a person to the degree of science, and then disappeared.

Will Zachariah be the only one to sell these strange hypotheses?

In fact, no, it was the era of the sixties and seventies that was marked by the fashion for an unsurpassed system of hypotheses of aliens who flew from the sky and helped people build their civilization, and then they left and left behind some of the effects of aliens that indicate their presence.

According to some hypotheses related to the planet Vuitton , it had a climate similar to our own. water, air and light, and was inhabited by creatures sane and refined and outwardly similar to humans to a large extent, but the idea of the inner courtyard of the planet and developed civilization in its entirety, leaving no survivors behind, the idea of a level of quality, therefore, supposedly believers of this theory believe that the inhabitants of the planet Vuitton fled from the dying planet and came to earth and played a significant role in the transfer of civilization to people, but they left the Earth for a reason, making a person to a certain extent

But why don't we see pictures or skeletons of these aliens?

Fitton : the mystery of the lost planet
The mysterious routes of pills
In fact, there are many effects in the world left by us from past civilizations, everyone talks about people who descended from heaven and were enriched by people. Including, but not limited to, the Dropa Stones of the ancient Chinese, and if we believe that he tells the Chinese that they broke the cipher on these mysterious translation panels, then the story engraved on it tells of a Chinese tribe. who came to their ancestors from space.

There is also a fee for the walls of the desert caves (Tassili) in Algeria, showing us people dressed in what looked like space suits, and this fee is estimated at 10,000 years.

The site provides Toro were (Toro muerto) in a beer, which contains many figures similar to people wearing what looked like space suits. We also have a board of caves and Indiana Vandzhina (Wandjina) in Australia in the inscriptions of people dressed in something like the space suits, breathing life back into these fees to more than 4000 years.

Fitton : the mystery of the lost planet
Some of the inscriptions in this cave are mysterious
There are also riddles of the pyramids and the civilization of the pharaohs, riddles and maps of Baghdad, riddles and exotic engravings of some ancient civilizations of America, like the Mayan civilization, where there is a temple inscription, Someone similar in the cockpit of a spaceship and holding some kind of device in his hands almost coincides with just one Sumerian goddess, is this a coincidence ?!

There are many - many such relics of a mysterious other, about which some in the place of a nightmare had never even had to talk about, perhaps even in the articles to come.

What do scientists say?
To be honest, most scientists deny what is contained in this story and characters, and they are just stupid, astronomers in particular do not believe that there was never a word about the civilization of the lost planet, but that many of them doubted that the planet

was originally missing , even iPod's law of distances between planets is no longer believed by most scientists, despite its oddity, but this does not mean there is no mystery or mystery surrounding our solar system, scientists complain that there are actually other planets in our solar system. but they are too far away so that they are difficult to find.

What about the mystery of the asteroid belt between Mars and the Moon?

In 1980, the number of asteroids that were discovered and determined their orbits around the Sun was 2,289 asteroids, and today we know that their number reaches several million! About 1.7 million of them with a diameter of up to a kilometer or a little more, there are about 200 asteroids with a diameter of more than 100 kilometers, including the asteroid Kira, an old drop of 670 kilometers, that is, if its surface area the total area of Iraq is approx .

Some scientists who support the theory of the lost planet argue that the source of these asteroids is a planet that exploded between Mars and Jupiter. The spacecraft of the American Voyager-2, which penetrated this belt on its way to the buyer, recorded the existence of asteroids with a diameter of twenty centimeters or less.

Asteroids have an irregular shape, not spherical, as a result of the weakness of gravity on them. The weakness of gravity leads to the escape of atoms and molecules from the gases of these asteroids, and therefore without air blankets. As a result of their great distance from the Sun, these bodies are cold, with temperatures on average about seventy degrees below zero. Asteroids move in orbits that are needed.

It is worth noting that not all scientists believe that the asteroid belt arose as a result of the collision of two planets. Currently, there is a basic scientific view that this belt is the source of the big bang that originally occurred in our solar system, and that these asteroids are stuck in this place due to Jupiter's strong gravity.

((Cicada 3301)) Organization of Geniuses

Cicada "Cicada" This name may seem strange to someone at first glance, but after studying biology, you would think that the insects called the cicada "Cicada" yes, these are cicada insects, but our topic today is not in biology, but in the group that appeared on January 4 2012 insects, taken from the name of the shape of these insects, the symbol in their message and there are no reasons for the unknown about their choice, I do not force you, dear reader, to guess about the reason to call them geniuses, but it is normal to talk about geniuses after you find out, what they did to the minds of the witnesses of their message, what they did to the minds of the witnesses of their message, you will someday read, Dear Reader, make you feel like you are reading some kind of science fiction and it could actually happen.

((Cicada 3301)) organizational genius

First, let's talk a little about the band's origins.

Cicada "Cicada" organization does not know anyone on this planet, who it belonged to and who is its patron or to prevent surveillance or even if it were independent, no one knows, except for them and they hear speculation about it ..

The category of people accused of "NSA" The US National Security Agency behind it, the class says the area belongs to one of the banks that are seeking geniuses and elites of cryptographic experts to work on a system to protect banks and funds in light of the large number of hacks , which lead to economic crises around the world and the collapse of many companies due to these breakthroughs, and the category blaming the organization of Freemasonry for creating the cicada goals are outlined in the diagram and designated as one of the teams, Pirates of electronic "hackers" form a group of geniuses in this field for the digital world ... Opinions and statements and frequent controversies, and no one could prove any of these charges.

But in my opinion, the first and not the second and not the third do not obey, I think that this is a team of pirates "hackers" and for some reason it was founded, and therefore the dispose method that they follow their Twitter account to an unprecedented degree , "twitter" is completely anonymous and untraceable. email and Twitter accounts, more accounts are being tracked.

((Cicada 3301)) organizational genius

The beginning of the puzzle picture one
On January 4, 2012, the first letter of this product appeared allegedly on the 4chan website, where the letter said that the organization was looking for members of the Council of High Performance and Intelligence, and they stated that their message containing an encrypted message would be the few who could find the beginning of the road. to get to the area looking to meet these elites who can decipher the code.

Hello.

We are looking for highly intelligent people.

To find them, we developed a test.

There is a message hidden in the image.

Find him, and he will lead you to the road that leads to us.

We look forward to meeting the few who can go all this way.

Good luck.

Hello.

I am looking for such smart people.

To find them, we are checking.

There is a message hidden in this picture.

Grandpa to suggest in the trail that connects you to us.

We look forward to meeting the few who can access the final stages.

Good luck

The message was kept in the form of a cicada, and has become the goal of many people around the world who are interested in joining this organization.

The image spread on the World Wide Web as quickly as light spread to everyone's concern.

Hundreds of crypto experts and hobbyists, cybersecurity experts and hackers, "hackers" in this world are looking for it, and many, if not most, cannot miss the first challenge, why? Is it possible that there were no geniuses? Only the members of the group who give this test?

90% of those who participated in this test, they said that the blade is not removable and it is a lie and is put for attention and glory, but some of the category is a little that is left on her mind that this code is guaranteed and can be displaced and out of tried all the strength to even go through one of them the first stage even brought back the debate about her again, but then what? After the group posted on their Twitter account, "Twitter" prompts help for the participants, and then returns to class who is the chef to get a solution, but there is no way only hints and baptized the group posted a hint every January each month years since its inception on his Twitter account to begin a new phase of wandering and mystery in the search for a solution to this puzzle, so that some Members have formed teams and groups to collaborate in solving this problem.

((Cicada 3301)) organizational genius
Hint letter
For expert advice and e-security and pirates, January 5 was the date of their annual attempt to decipher the puzzle, the day featured images from the mysterious Twitter region:

Hello.

Let inspiration be your ally as you begin your pilgrimage to where the lights went out.

3301

Good luck

Hello.

An insight has descended on you, your pilgrimage has begun. Enlightenment awaits.

3301

Good luck

The picture is a symbol of a cicada insect, in fact, the background image is not black, but dark gray, and if I open the image with a program to improve the sound and increase the level of shadows, then the icon shows the assembly: cicada

Even if the edited image, you will not find, in which only the area code and nothing in this case is indicative, but this confirms that the picture was made under the hands of the members of the organization, but does not have that the photographs carry the message of another hidden, this encoded message is not just toys, they are needed to achieve a complex way to extract the hidden message and tired of the human mind, it needs a special kind of programming.

After analyzing the Audio Recording, you will receive a quote from Ralph Waldo Emerson's article name: "self-reliance" and a number of numbers and encryption mechanism

In a sense, both the number hidden in the picture refers to a paragraph, that is, to a sentence, and the word freedom in Emerson's original, composed by these symbols, can end up on the title website

At the previous stage, everyone managed to find the solution to the site in the dark web, there was another riddle from the riddles, the organization of the cicada.

The third puzzle "mystery of the duck"

((Cicada 3301)) organizational genius
Accessing the online website The Dark ..... Image of a duck
Surprisingly, the site had a picture of a yellow belly with the words "Oh, I got trapped ... you can't seem to access the message."

Do not discourage the researchers, there may be ducks and what is written on them with a hint, nothing more, try the researchers of image analysis like its predecessor.

Another surprise:

There was another message hidden there, he brought them to a famous place for those interested in technology.

On the forum, she had codes-complex symbols used by the Mayan civilization, decoding, another code appeared.

"Until now, no codes were required, and I have no coding skills, but he started it in his post."

((Cicada 3301)) organizational genius
Thanks for calling ... mysterious phone number
At the previous stage, the researchers analyzed the symbols taken from the Mayan civilization, and their leaders encoded them in turn into a book of poetry telling the story of the receipt by King Arthur of England of texts of the local religious province of Wales, dating back to the pre-Christian period, and from there into a poem of the electronic publication of the poet William Gibson's 1992 "Flaubert Disk" software to erase the contents after reading a poem only once.

And between each phase and stage, and hours of painstaking efforts to decipher the riddles.

Here thousands of programmers began to jointly solve puzzles on the forums of well-known point owners, anarchists and hackers.

One of them, I knew that analyzing the code hidden in the poem would lead to the following message: "contact number" 0000-000-000 "is dancing fake.

A Texas phone number is called an automated message asking you to go back to step one and extract prime numbers from the original image.

By multiplying the numbers together, the researchers discovered that the number new Initial is actually the name of the website 00000000.com (this is a fake), where the hour several descendants and the cicada code appeared.

((Cicada 3301)) organizational genius
First ... and under the codes of the Mayan civilization
When it arrived, the countdown went to zero, I selected the first one and showed the geographic GPS navigator settings replaced with symbols for 14 locations around the world: Warsaw, Paris, Seattle, Arrivals, Arizona, California, New Orleans, Miami, Hawaii and Sydney ...

Help thousands of researchers with the "GPS" setting conditions, and there they found that the pictures hanging on the lamp posts show you the cicada symbol, the Chevron visually "QR" and are a black and white code designed to be read by a cell phone camera and guided you to the site, and from there to the location of the cobweb hidden in the dark.

((Cicada 3301)) organizational genius
QR code .... which I found on the lamp posts
After visiting the site of a limited number of people, I was shown the following message: "We want the best, not the belonging."

Here the story of each deployment of the solution ends and each of us did the work on the solution himself, it was one of "ours" in the perspective of the cicada, but what happened to those lucky ones who got to the site by exception? Or is it the end for everyone? And we are turning to the world of assumptions, not facts.

But what if you knew, dear reader, that the riddle has already been solved?

And everything that I read under these words was just the disordered behavior of researchers who led them into a trap prepared by cicadas tightly, do not rush, I already told you in the first lines that they are geniuses in chess and counting them several steps ahead is a distinct style that makes them winners in every game.

Back to the duck puzzle

Another letter, a puzzle and another and new stage, but this time with me, in cooperation with cybersecurity specialists, the doubt began to change inside me, why do you put all these tests to the cicada in order to finally solve for the participants: "we want the best, not the best "?

Is this really the end?

Something inside my head tells me, "You are not from the sight."

The deadlines for that one of the phases, containing one more letter indicating the correct path, do not fall into the position in which he wrote to us, we want the best, not followers.

Why else say that we want the best?

I went back to the first phase started again, but the image of a duck of a different nature from all the puzzles, it was just an image containing a message, like the previous ones from encrypted images, but it was something that didn't seem to notice it from the link of this picture , I can access this image follow the principle of decoding the previous images, the side you put the error code all the code from the code it! Isn't it weird that they chose "duck" this time?

And does not contain any part of the images of insects?

Cicada will not follow the same method throughout the code

It is not strange to change the encryption method every time it testifies to their genius and the rhythm of the participants in the maze there is no way out, they are looking for the best, not connected with it, any person up can happen for his beautiful, it is not reasonable to be this is the end, dog up or not up.

But I know that the other meaning and the other code are completely different from the one found, maybe because I did not want to be a follower, like others before me, this time the cicada put a message in the test, one of which shows the path is wrong, and the other - right.

Lead me to an analytical conclusion, completely different from the one from which it was received by the one who was accepted and agreed by the method of analysis with the previous ones, I arrived at the place of the hidden letter, and then wrote a letter in which several lines contained numbers and a few words, in which he wrote: "You know what to do, good luck."

Hi you know what to do

Good luck

one

2

3

four

3301

The numbers from one to four are the number of lines of those numbers.

I did not enter the numbers, you cannot display a special message. Just numbers, to be honest, I didn't know what to do, are you making an equation out of them? But she was larger than the rest of the pictures for work, and after she analyzed and manipulated the photo editor, she discovered in the depths of the picture, a person and his numbers on the side, and the cicada people have above their heads, and there is not a single number letter in her file.

Anyone who sees people may think that it is very simple, but it was even more difficult for me than ever when I had these bright ideas in my head.

Why was Cicada included in the 3301 dance after this name?

Since something was deduced from the name of the cicada and 3301 that was not taken into account, it was not only a big riddle from which passed through my head that I can decipher from them a number similar to one of the numbers that existed in the depths of the image on the side of a person , as well as as a confirmation and a hint of a certain number of them, I am sure that other conclusions of the finds, but I did not disclose this to anyone.

It was there that at the exit the riddle of numbers was solved, which came after decoding the picture of the duck, using prime numbers in the original image and processing was formed using numbers in the message and the name and number 3301 were displayed

To get me to output to a phone number other than the one you got when they kiss, but that phone number is correct, which made me make sure it was needed in her last 3301 when contact answers, autoresponders, with numbers denoting the web -site looks like a link no researchers have deciphered the poem yet, but different in name, and when you open it to find where the insect symbol and sound clip are, but it's just music, no words needed , just played the famous "Swan Lake - Swan Lake "for Kaikovsky, gentlemen, 4 years of work on deciphering this code, I got pains in my head, and every time I see that I enter the labyrinth in a completely different way, so I decided to either finish and go crazy, or leave everything this, but hey, curiosity and scientific potential prevent me from leaving, and if at the expense of your health, then I don't know the impossible and after all this progress you left it as if I had a free passage?

((Cicada 3301)) organizational genius
Analysis of audio recording
This time I used the software analysis of acoustics, it was not so difficult where it was necessary to sound a deep wave, confuse melodies and music, a number was guaranteed, and when I edited a photo as a digital file, I found several symbols large and small among millions of digits, collected letters in order and distributed the numbers between them, put the numbers in order after each character, from behind the name already led me to the site in the dark web, then to the puzzle, when I saw it at first glance, I almost went crazy and passed out, maybe maybe because I was able to decipher them. or because I saw that

the code could place a creature from outside this area. - You imagine, dear reader, that you will put yourself in my place "

Everything that I found in the picture of an insect is drawn with numbers and symbols, and symbols, and not its image was written. You can copy and paste them in any editor. And it is written what is written in the tooltip that appeared after the first puzzle "" let it be an inspiration for your ally. Their pilgrimage began to where the lights were burning. "

3301

Hello.

An epiphany has descended on you. Your pilgrimage has begun. Enlightenment awaits.

3301

This is a cicada in the process of the last gentleman, this is what I learned after my long-term research about her life, I asked myself, is this the end of the test for me?

No, it wasn't over yet

He used the solutions of a former who was infatuated and believed that the regime was using reverse engineering of "six decimal places" to solve the mystery of the latter to wake me up to infinite madness, he was just experimenting, maybe in detecting an error message he hinted to complete the road and this is what happened.

((Cicada 3301)) organizational genius
This picture is not real
With the help of a cybersecurity expert, there was access to the solution to the puzzle after I show you what I came up with and that hit us to direct us to the location of the wave on the upper layer of the Internet "Clearnet" the puzzle is thinner than the neck, this time to me nothing is needed from the name or shape of the cicada and even the dance 3301, just the numbers a million and some symbols are written and laid out carefully, forming a structure reminiscent of the buildings of the Mayan civilization, and a message above that said: "We can hardly wait for this moment."

And some of the deals with her are true ... that's the end, dear readers.

I started to wonder if you eat my mind will you be exclusive? And will I get myself a follower?

It was the end for me, I don't know the beginning of the end.

I don't know where I will start, from the first or from the middle or from the last section

This is what you need for a hint, to solve it, losing hope, we are waiting for a hint as soon as possible

In fact, the promised every year in the month of January, but no luck, and a few days ago from the writing of this article, I specifically found a hint that pushed me to the fact that I found it in a place not in the mind of the devil of ingenuity, so how will the mind be man? The clue was not published anywhere, it was within the puzzle, a few symbols showed me in the first place and how it would end

I found it by accident, Gentlemen of the game good luck - play with me this time, not with my mind, I found it, gentlemen and took months apart so that the code saw the internet world, unblocked it and got an output to my email address to connect with them, everything that we went through and all the time was lost for the email address, blimi, the atrocity of the son of Adam.

But the point is, after all these efforts, is it worth reaching out with you? Who are they? What awaits me? I didn't expect this to end

What do you think? Doesn't he only exist in science fiction films or science fiction?

This time it is not a science fiction movie or a novel, but the fact that I lived it, I made it the greatest novel and a fitting science fiction movie.

In the end, I apologize for the attached sound that got to it and the titles that gave me a ride on the rest of the tests, it cannot be written about the episode in its entirety, and it shortens and explains some of the steps incompletely because the topic talks about this organization and the experience of the participants in general and my experience and simple explanation in both cases, and is not intended to suggest solutions, but some of what I have experienced and others, and dear reader that the mapping of stages in my experience is part of this and is not complete, I wrote some steps related to each other and some programmers and information security specialists, and some steps that got me on the right path, and the final steps, and what width of the image within the pre-existing A available to the general public and fake images, audio and telephone numbers for NASA implementation, and the addresses of the real ones should not be published, especially what she got from my experience. And that's because I don't want anyone to follow me, every human path.

The ocean suddenly disappeared in Brazil ?!

Since a strange fish phenomenon spread on the coast of Brazil and Uruguay on the afternoon of August 11 this year, which may soon be resolved in these areas, the water level in the ocean has suddenly dropped to serious levels without warning, prompting the authorities of these states to take appropriate measures to catastrophically close these shores.

The question scientists are now asking is why this phenomenon happened so strangely?

Understanding premise:
1 - global warming

The ocean suddenly disappeared in Brazil ?!
The first has turned into a barren desert!
High temperatures in recent decades have led to a decrease in the water level of the oceans and seas on the coasts of many states, but this does not explain such a sudden change in the level overnight, there is a hypothesis that the melting of snow in Antarctica as a result of global warming can lower sea levels, because salts, which are present in the snow when it melts, can draw out a huge amount of money for the density level of the natural ocean.

The ocean suddenly disappeared in Brazil ?!
Glacier color refers to the amount and type of salt it contains

However, the most important thing that darkened this group of defects is why you did not lower the water level in front of the African coast on the other side of the ocean, corresponding to the shores of Brazil.

2 - sea currents

The ocean suddenly disappeared in Brazil ?!
Is the ocean currents the reason?
Sea currents in Brazil take water and distribute it to Antarctica and Africa as a result of different navigation, the intensity of these currents can vary, which leads to the diversion of water to the shores of others.

As we can see here, a stream is a stream that may be responsible for pulling water out of the shoreline due to changes in temperature and salinity.

3 - an earthquake underground at the bottom of the ocean, this is very likely to cause an earthquake with the removal of a huge amount of water from the surface, and then it is reflected by a second, which leads to a tsunami wave.

The ocean suddenly disappeared in Brazil ?!
Significant decrease in water level
4-hurricanes

May was baptized by hurricanes to pull huge amounts of water from the surface if we see a shift in the position of Uruguay of our people as the region that produces many of the hurricanes that hit Antarctica.

Finally, the question remains open about the phenomenon, which haunts scientists, as well as the authorities of many countries about the consequences of the disaster and the potential of this phenomenon.

What do you think, dear reader? Do you have an explanation why this is happening?

Each point is a galaxy.

There are many types of horror and its causes, for a moment some may be surprised to ask about such a topic and how his relationship was scared and terrified, but believe that this topic is a scary fact about the universe we live in and about the nature of this magnificent universe.

...

Every dot is a galaxy
Fob things giant waves there are many people
Psychologists talk about fear of giant things and call these fears by the term (megalophobia) Meg Tuvalu, or fear of big things. This fear is experienced by many people when they see something supernatural, such as an organism, a giant or even a tall building, or maybe when they encounter the inside of a very spacious hall.

But hey, I guess don't even say that our topic today is the biggest of these issues. Because even the words "big" and "giant" shrink and become meaningless when it comes to what we are about to talk about.

...

Let's articles, in short, our today's fairy tale is a picture!

Yes, just one image.

I'll put it at the end of the article. A simple image is small dots On a black background! It's that simple!

But make sure that if I didn't scare you with this picture and didn't stir your imagination, and didn't change your way of thinking, and didn't pick it up in your mind, then you definitely didn't deserve it.

...

** The introduction should be:*
There is a phrase that people always say, when you add your lifestyle to one of the places, you must have heard that "Allah's land is wide."

This is true if the planet is very wide and more than 7 billion people live on its surface and its number does not have an infinite number of living beings and creatures.

** Solar system.*
We know that the planets revolve around the Sun and this solar system complex. Now let's compare the size of our earth clearly with the size of the planets in our solar system, take a look at the photo of the size of Jupiter (Jupiter) compared to others, (Earth) and ride !!

Every dot is a galaxy
What if you compare the volume of these planets with the size of the Sun (Sun), note the size of the Earth and travel.

Every dot is a galaxy
Can I see the Earth in the previous picture when you put them next to the Sun?

But is the sun really big, let's compare the size of a small star close to the star of humans (Sirius) like the giant star Arcturus (Arcturus)

Every dot is a galaxy
But is the star Arcturus considered a giant, in fact, let's compare the size with the size of the star Antares (Antares), note that the arrow indicates the size of the Sun-Sun has become just a point that cannot be seen with the naked eye compared to such giants !!

Every dot is a galaxy
** Tip*
Collect the stars and destroy from the inside Kiev of enormous dimensions, which we call the "council" spiral, a huge one consists of billions of stars with its solar range containing what could not be limited by planets.

And if we compare the sizes of the giant stars, the seventh largest in the council, they simply turn into dust aerosol atoms and are nothing compared to the volume of the council itself.

Note: If the sheer number of stars that we know and observe in the sky represents a small area of our galaxy and council chamber, then the little yellow color shown in the image contains everything that we see from the stars in the sky:

Every dot is a galaxy
All the stars that you see in the sky at night do not appear only in the yellow circle shown in the picture
** Each point is a galaxy, complete:*
In 1986, scientists from the European Space Agency (ESA) managed to capture an image of what lies outside our galaxy. pointing the telescope through the vagina into a small space called the lockman hole and this image:

Every dot is a galaxy
The previous image is a fractional listing of the universe, and the entire small dot in the picture is a "giant galaxy" containing billions of stars.

Mysterious creature and unknown assassin Chupacabra

The legend of the strange and the mystery of the confusing .. Chupacabra "Chupacabra" is one of the myths of the modern best broker in the north of North America and the south, claiming that the creature is a vampire, roaming the villages and countryside and killing livestock, leaving the body intact without defects, incomplete drinking her blood, and that's what caused the buzz and focus on what might be the memory of the serial killer's mystery.

The most common description is that the object resembles a reptile, with the crusts or scales changing to a gray lawn and sharp on the back like feathers. He has red eyes and sharp claws, since he looks like a kangaroo beast, they should be short.

In the mid-nineties, on a small island called Puerto Rico, and this is one of the Craig Islands, and it is in the center of the canvas, a legend spread about a strange and mysterious secret about a predator whose purpose was not to kill all those animals that ate. all the bodies of water that killed her and were found as they are without physical harm multi.

Mysterious creature and fight an unknown Chupacabra
The bodies of the killed animals did not contain were almost intact
But two holes were found on the neck of each victim, it is strange that her body was completely devoid of blood, but it was dry, traces of fangs are visible on the bodies of killed animals ... this is a really difficult question, any of the predators that we know, he attacks its victims to consume their blood? Probably not, and the population became in the face of a strange beast, the identity of an unknown person, and became a mystery to them.

The terror began to spread to regions and other countries such as the United States, Chile, Brazil, Mexico, Puerto Rico, and spread fear among people, especially farm and livestock owners, simply because the creature focused on farm animals.

Start of attacks

The attacks started violently, almost every day, a large number of farm animals were killed and I found teeth marks on the neck of the victims, and all the animals were almost killed in the same way.

In 1995, about 8 carcasses of slaughtered sheep were found in Puerto Rico, and these attacks never stopped, but arrived in Calama in Chile, where the monster attacked one of the animals of the "llama", which is an animal that looks much like financial terms of the transaction and form.

The villagers were attacked at night, and as happens every time, there was a hole in the throat and drank the victim's blood in full. the wound was very transparent ... one could also look into the victim's body.

He claimed that one of the villagers saw the creature ambiguous, he said that he was almost like a small bear and its color tends to gray-black, the witness added, saying that the creature has a series of spines starting almost from the front to the back of the tail. Witnesses said that he had two small hands with three claws, and that he looked intimidating because he stood on his hind legs like a kangaroo .. The creature killed about 188 animals. And of course, this figure is only in one area, with the exception of regions and other countries. We also included dogs and cats in the list of animals. When investigators began and some doctors began to investigate the case, the timing of the attack changed slightly, and also deployed part of a security patrol overnight in an attempt to catch the Chupacabra monster.

Mysterious creature and fight an unknown Chupacabra
They called him a scary creature with a row of thorns on his back.
Again comes the mysterious monster called "Chupacabra", violently bombing, stepping over the line, this time attacking a woman on a farm in Puerto Rico, it was said that he tasted her blood, which means that the person is also on the list of the Chupacabra predator. doesn't seem to be slow to attack humans.

He was also attacked in the city of the IOC, a small town where the owners were engaged in agriculture and cattle breeding, the monster attacked one sheep, and also attacked a guard dog too. it was a German shepherd and his dog, guarding the sheep, proved that the dog deserved it, guarding the herd of wolves, but the arrival of the beast did not say a word to do something, but he himself was one of the victims of the Chupacabra.

Trade police stations near villages and village new complaints about the death of a cow, she was found dead in the morning, and the question is, how can a creature with an elf-like chupacabra defeat a huge cow! -from? We know that all the predators that attack livestock and attack the massif organization, but the Chupacabra was alone here ..!

The doctors realized that they had a mysterious creature in front of them, and did not have any details about what it was, but they described this creature as if a strange breed of wild dogs could be born from interbreeding between dogs. They said that this type of dog looks like a reptile.

Similar attacks began to occur in Argentina, Brazil, Peru and the Dominican Republic.

The stranger's paradox
Mysterious creature and fight an unknown Chupacabra
This creature is smarter than any other animal.
1-through the security patrols that were going to look for the creature, its attacks automatically stop, but it returns again after the security forces and the sea leave. This means that the creature is very sophisticated, and rubs its intelligence into the intelligence of any animal. In See, a rare creature, the man says hopefully he can jump about 20 feet, the equivalent of 6 meters, this becomes the Chupacabra, the world's tallest predator jump.

2 - in the description of the strange, it is also said by one of the eyewitnesses, the creatures emit a pungent smell, similar to the smell of sulfur.

The 3rd cover research team was unable to explain what is going on through attributing the myth to some dogs with rabies, but the Chupacabra is not a dog in the original.

4 - the spread of attacks also in Spain and in some European countries, the strangest thing is how to turn the Chupacabra from America to Europe ..!

5. The team was unable to find any traces in some cases, despite the fact that the bodies of the animals were scattered all over the place.

6. Prove to researchers and scientists that the Chupacabra creature is very intelligent and has a sophisticated mind, unlike any other animal, because all the places that attacked it were well guarded by walls, cantilever buffers of closed doors, how could the Chupacabra open them !. ..

Chupacabra encounters with humans
Mysterious creature and fight an unknown Chupacabra
Feel free to attack people
Joy Royal says: "I was in a state of panic and fear near my house. Next to some chicken and your family of mine I felt bad and strange, and unfortunately I saw a monster with two red eyes, scary two, he had no ears, he stood near the door of the house and wanted to leave, and he looked not scary, disgusting, I think he was a demon or an alien, but he left soon after. "

Evelyn Gaspard says:

"One night I woke up from a dream and wanted to spend my need, the bathroom was separated from the house, the atmosphere was very calm, and the cool breeze blew slowly, but suddenly something fell on my shoulder, it was heavy, and when he turned and found monster so ugly .. and the teeth are big, and the scary creature emits an unpleasant smell .. attacked me and sank his teeth into my body, and then after that I lost consciousness, and so it went on until I woke up in the morning and found her husband next to me. " Evelina has this creature with three stores, and some of the specs, which

were coincidence, were made by other eyewitnesses, which means that the legend of the Chupacabra may have been real and dangerous.

Jack says to Glenn Minecraft:

"I did not believe in the real presence of the creature, the Chupacabra, but now I am confident that what I have seen before is not a dog or any other animal." I wrote a press and media article on July 11th.

In September 2009, a CNN TV report aired on the network, showing a video of the close proximity to the body of an unknown animal. They also report that local residents suspect that it may be a Chupacabra. After reporting on the body of a former student mummified from Blanco, Texas, he received the body of a former student, he said that he soon found this animal in his barn, where he died from a poison that was developed for rodents .. the said scarecrow believed that this coyote was genetically excited.

On September 18 of the same year, he sold taxidermist Jerry Eyre from Texas, and gave the body to the Museum of the Lost World, as reported in the "Syracuse Standard", on September 26, 2009 the museum presented the creature to its people.

Origin most likely according to legend
Mysterious creature and fight an unknown Chupacabra
One of the explanations for this creature at the level of objects of science fiction. Benjamin Radford's exploration of the hidden animal world lasted five years, as documented in his 2011 book Tracking the Chupacabra, which concluded that the eyewitness account of the first and original Puerto Rican Madeleine Tolentino creature level "in a horror film and science fiction" kind. "

To match the description of the creature, the torrent almost described Tolentino, and they watched the movie before submitting the statement: "It was a Chupacabra-like creature with spikes on its back," Tolentino said.

Finding Radford after searching for him in Tolentino thought that the beings and events she saw in the movie "Views" may have happened in reality in Puerto Rico at the time, thus summing up the most important description that cannot be trusted. It is believed that Radford thereby undermines the credibility of the Chupacabra as real. In addition, the anatomy of some of the bodies is in none of the blood source reports, while the veterinarian's analysis of 300 casualties they did not fully download.

Divided Radford's reports into two categories:

1 reports from Puerto Rico and Latin America where animals were attacked and allegedly sucked on her blood.

2. Reports in the United States of mammals and most dogs and animals infected with coyotes with scabies, which people call "Chupacabra" because of their normal appearance.

At the end of October 2010, a biologist from the University of Michigan, "Barry Alone", to all reports of Chupacabras in the United States, simply replied with coyotes-coyotes, which would explain most of the peculiarities of this case: "when there is little fur, and the skin is thick and smells like rot." The one that attacks on goats happened because these predators are weak and no longer based on a formula, so you are forced to attack the past because it is easier than a rabbit or a deer.

Although many witnesses concluded that the attacks could not have been carried out by the dog or coyote because they did not eat the victim, this conclusion is incorrect. It can happen that dogs and coyotes kill and eat prey, either because they are not the same experience or as a result of injury or difficulty in killing the victim. It remains a prey for survivors after an attack, and then dies from internal bleeding or shock in the circulation.

And add in conclusion that the chupacabra is one of the creations of Krypton, Renato is sure that he knew hidden animals that have yet to be discovered, which still constitute a mystery to most researchers.

What's at the edge of the universe?

Since the foot looks up to the sky and ponders how the universe and its stars and planets are different, perhaps the first of these questions was about the nature of this universe and what to do if there are no definite limits, or rather, where ultimately to be us? what is there in the end?

What's at the edge of the universe?
It was the belief that the earth is the center of the universe.
I thought by the ancient Greeks that the Earth is the center of the universe and that everything in the sky revolves around them, until the Polish astronomer Joseph Copernicus (1473_1580) proved that the Sun is the center of the universe and other bodies revolve around it, including our planet Earth.

Are we really aware of our size in this universe, or, in other words, are you really aware of the immensity of this universe that surrounds us?

For several centuries I think that most people, including astronomers to our solar system, where the Earth is the entire universe, even the back physicist "Isaac Newton (1642_1727), and he said:" that the solar system is only a small neighborhood in the Great Galaxy called the Milky Way, and in turn that is the universe.

What's at the edge of the universe?
Astronomer Edwin Hubble
This was a widespread belief until the middle of the twentieth century, when the American Edwin Hubble (1889_1953) entered the world of astronomy and made the last real revolution in his discovery that our galaxy, the Milky Way, is just a drop in a sea of galaxies, there are no similar ones, their number is numbered billions, they are such that our view is radically approaching the size of this universe that contains us, but Hubble did not stop there, and also proved that the universe is constantly expanding, that is, that everything in the universe, from galaxies and stars, is moving away from each other friend, which brings us to the question we asked at the beginning of this article:

- Is the universe ending? And if the latter depends on the extent to which you arrive? "

What's at the edge of the universe?
The Big Bang Theory explains how the universe we know today came into being!
The proposed theory of the origin of the universe or the big bang, "Big Benga", that the universe was at a minimum point of zero, but it was crushed by a strong cluster, led to an explosion to edit most of the money that we know today and which formed the universe, led gravity to the interdependence of the parts of this article is the problem of stars and galaxies, but we ourselves are humans.

Scientists believe today that the Universe will continue to grow until it reaches the degree of realization of this energy, which arose as a result of the Big Bang, which logically leads to the compression of the Universe and its return to the point of origin, i.e., that the Universe will collapse by itself yourself in the process of crushing high for all the money you put in.

But wait, the plan of modern science has proven that the universe is not consistent, but it is accelerating, that is, the speed of this sequence continues to increase.

I was confused by this fact of scientists, the laws of physics that we know say that everything is deterministic, it must contract in the end, it is impossible to continue this funding only if there is another force of the mysterious work of the opposite action of gravity, called dark energy.

Now, can we truly grasp where you belong to the boundaries of our universe?

So come a little closer, understand this, let's first know the huge distances that separate the galaxies and stars that make up this universe, which our minds cannot easily avoid, this distance is considered huge compared to its accepted standards, so astronomers have developed completely different standards for measuring these distances, the values in it do not work, so they calculate the speed of light, that is, the distance traveled by the latter during the whole year, they call this optical unit, and the knowledge that the speed of light is constant is always almost 300 thousand km / s, The process is simple, we find that light cuts about nine billion kilometers a year.

What's at the edge of the universe?
Centaurs are mostly the closest star to Earth After company
And to give an idea of the dimensions of the distances in the universe, let's take the closest star to us after the sun, of course, "Centaur Sr.", which is about eight light years from Earth, that is, it takes light eight years to reach the earth.

Imagine with me, dear reader, a star and it is about a million light years away, after all this, you might think that it is impossible to determine the end of the universe, but scientists have tried to put what looks like a map of our universe, the parts that allow the universe to see light about 16 million light-years away, which means the maximum amplitude we can see through light is called the "horizon of the universe", but you might ask what is behind that point called the horizon of the universe, right?

The answer may sound complicated and complex, but it can also be very simple: Scientists say that we will not find anything other than most of the universe, because the universe is just so light that I can use our technology in monitoring to see it, but we know that this technique is still considered primitive if we use it on a scale of this size ..

At the end of the day, this journey sounds like we haven't received an answer to the question we started with: "what's at the edge of the universe?"

Perhaps the strangest theory to offer a solution to this question is the multiverse theory, which has emerged in recent years and is accepted by a number of astronomers and physicists.

What's at the edge of the universe?
Are we the only ones living on this planet, or are there other factors?
To say that this theory is that we do not actually live in the same universe, but if there are multiple parallel universes similar to our universe, then this theory is that the Big Bang that we mentioned earlier could and not be a unique event, but it has other explosions

that led to the creation of universes, many of which are similar to our universe, and yet these explosions continue to this day.

And the strangest thing is that this theory assumes that each of us, humans, is duplicated in other universes, in other words, maybe for you, dear reader, and you demand to copy this article, how you like fishing or playing football in another universe etc...

Indeed, the idea may seem strange, but, crazy or hard to believe, however, that scientists while working on the study of these theories say that they may be able to prove these hypotheses, based on the accounts of sports physical, it is very difficult to implement them here in detail. ...

Finally, but not least, this is due to scientific and technological progress in various fields, it is not surprising that the correspondence of our scientific discoveries and facts can reduce our view of the universe as we know now, this is Aristotle to Einstein and Galileo and Newton and the end of Stephen Hawking .. the shades of scientists of that era are replaced by others and change our perception of this mysterious and exciting creature that we call the "Universe."

Lake Tahoe Tse monster

Since the time of the itch dates back to the middle of the nineteenth century, members of the rat tribes and the Piot Indians betray the legends of the miraculous and exciting settlers of Lake Tahoe of America, where they thought that a monster from the water taken from the depths of this lake, we caught it, and tell their stories that the creature exists in an underwater tunnel called the cave of the rock, nicknamed by the population already by the name "Taho Tessi", the proportion of Lake Tahoe is such that I continued to observe it until the modern era.

Lake Monster Tahoe Tessie
Lake Tahoe attracts tourists not only with its beauty, but also with the rumors that revolve around
Lake Tahoe is located at a distance of 1,879 meters at the top of the series in North America on the border of the states of California and Nevada, considered one of the most beautiful places in the world. Maybe what is causing the surprising attraction for those on the lake is the rumors that flow past the residents of the region in addition to the tourists available, and perhaps most notably the Tahoe Tessie mysterious creature.

According to legend, Tahoe Tessie is a creature of Christology creatures whose science is to study hidden animals such as dwarf miss and chance, Loch Ness monster, etc. compensation originated from the creature in North America's largest alpine lake, Lake Tahoe , and is due to the participation of a rare creature 10 to 20 feet long with a body similar to that of a snake, its color is between gray and black, its skin is usually smooth, plus it has the features of an animal and crawling.

Types of the alleged creature

Lake Monster Tahoe Tessie
Some claim to have seen a long snake swimming underwater!
In five budgets of the last century, police officials were out of their service, while they saw huge black money appear in Kiev, how they simply run through the increasing speed of their boats, and claimed that they exceeded 60 miles per hour.

Anonymous eyewitness from California says that he was with his friends in three small schools to learn how to catch salmon, as they saw the creature large, its features were unclear, and it was in the middle of the winter of 1979 in Homewood, the witness added that the creature was looks like a lamp post and dives in and out of the water for another moment, in addition to splashing water from the top of the head .. the witness and his colleagues were left stunned after seeing the device until it suddenly disappeared into the sea floor.

In eight hunter budgets, they see a snake 15 feet long, or about 4.6 meters long, which was cause under the surface of the water near a rock cave, and a few weeks later, two divers reported that they had found a cave under water by shooting an unknown creature, admitted that the creature had large fins, but they could not catch it.

Lake Monster Tahoe Tessie
Here the boat and Kai Green will soon disappear
Sometime in the late 1990s, a coach blasphemed a group called "planets" that he saw a glimpse of what looked like a boat and that Kai Greene - small, two people - soon disappeared into a sink of water when I approached the coach from that the area, where there were no traces of the kayak, is a form of a puzzle of what he saw that day.

In 2004, ostensibly as a waiter at a hotel near Lake Tahoe, who photographed something black in the water that supposedly is the head creature of Tahoe Tessie, I observed the same person - who was with his family this time - a creature similar to in appearance to a sturgeon with a white nose, but the creature moved up and down like mammals, and did not save it from the direction of the trend, like the rest of the reptile.

It is estimated that the Tahoe Tessie show in June of every year, despite reporting a lot about his participation throughout the year, here was curious about the interested monster legend about what to do if it hibernates or not? Although the weather in winter will be cold even on Lake Tahoe.

As for the authorities, they refuse to investigate what lies in this lake, add waiver of fines around it to give unimaginable compensation after scaring a monster, the population of Taho Tessi, and refuse to admit the existence of a local agency competent in biology. divisions. But reports and opinions, and even the missing people ran into denial of the existence of Tahoe Tessie.

The premise of understanding

Many of the theories surrounding the monstrous Lake Tahoe (tese) theme are the most popular:

Lake Monster Tahoe Tessie
Some said that this squad is just a kind of dinosaurs that lived millions of years ago.
The sea animal is a type of dinosaur species that lived on Earth millions of years ago, they are called by several names, including: Prior, Prior, Rector or OSCE, which makes this theory true, finding archaeologists at many sites for creatures of this variety found in the vicinity of the Nevada desert, as well as in the mountains of the series.

This theory is rejected by many researchers studying the problem of Tahoe Tessi, who claim that they rejected the cross-attribution of the formation of Lake Tahoe to the last ice age, many years after the extinction of these creatures. However, defending some of the people on this theory, arguing that these creatures can survive extinction by living in caves, isolated or caves, or even can adapt to the marine environment, as the rest of the amphibians did.

Another theory put forward that undermines confidence in the existence of the frightening lake monster is that all species and captured images are simply free-moving snakes or sturgeons, which in a huge structure will fancy people or make them think that there is a monster that inhabited the lake. ...

Lake Monster Tahoe Tessie
Images that prevent wearers from seeing the device do not appear clearly
There are several depictions of people claiming to be a Teese unit, but this is not considered conclusive evidence to the extent that it is decisive, they could be trees or glorious grasses cultivated brick floaters, fish large, ebb and flow, or even boat, clear in its characteristics.

There was an interesting incident on the Lake Monster that happened in 1970, when Jacques Cousteau led a French scientist in biology to investigate what lies deep in Lake Tahoe. Cousteau's debt in this campaign is something too scary to refuse disclosure, set his voyage of discovery, where the mysterious gas campaign concept came from in his utterance: "The world was not ready for what was acceptable in this lake!" the mystery that Cousteau never rubbed himself into any statements in connection with the most incredible event in the history of mankind.

Jean's puzzle: the cruel mother of Abkhazia!

The list of science is a set defined as (controversial science), or (notorious), or in other words, is not taken seriously, I just think that science is a fake, and shit and legends are the mean clothes of scientific methodology!

Among this list was an old note (alchemy), which deviated to the sea in a fairy tale was well rehearsed, in the eyes of science, for example, the search for the philosopher's stone (or the elixir of immortality) and a cheap metal to gold, and in our time the study of such sciences as: astrology, license, you are mine, alternative medicine, sings (are) on the list of pseudosciences that simply mean (science is asked to provide homework, on the contrary, no evidence can be examined by means of scientific methodology, traditional or updated, and to prove the authenticity of these facts and hypotheses).

In this mysterious world, the flag is controversial as it is interesting as a tonic for the imagination, and it taught (Christology) is a term that refers in short to the effort to find hidden objects!

Christology: creatures that we cannot see, but we believe in their existence!

Puzzle G: mother of cruel Abkhazia!
Cryptozoology: the science of finding objects that are not proof of its existence!
Consists of the name of the Christology of two syllables of the Greek language (crypto kryptos) and the hidden or invisible, as well as zoology تعنيوzoology, the general meaning of the term is (study of hidden animals) or invisible or known.

This science, which economic scientists consider to be a fake like myths and legends, tries to answer the traditional question: Are there any creatures on Earth known to man that have not yet been discovered ?!

Here we must distinguish between Christology in a broad sense, the idea of finding fractions and new species of animals, birds, insects, fish, etc., This last show is a systematic sound, all of it falls into the list of science (professional) recognized by all, the topic of today is about magical or mysterious creatures who celebrate their presence and consider it a fairy tale like: Bigfoot, Chupacabra and man (which implies his evolution and extinction in full) dinosaurs survivors!

As a science, its origins, although not caring, rid Christology of the fog of mythology, and not of the rules of the principle of "Bernard Home," the Franco-Belgian, who is rightfully the founder of Christology, red, and the book of the famous bear bears the name (at a distance of courtesy creatures), which was published by its original French edition in 1955 under the title (Sur la Piste des Bêtes Ignorées) And then I soon found great success and the book was translated and reprinted several times thereafter, the importance of this book is that it put the author in his place so that he would receive, without competition, a title (Father of Christology).

Puzzle G: brutal mother Abkhazia!
Bernard Home: First thing to do is find the hidden creatures of the note and put assets on it
But the book "Home" is not a guarantee of the knowledge of the newborn, recognized by scientists as traditional, because this flag has remained, until now, missing, therefore it is urgently required: scientific proof is concrete!

Regardless of the scientific and philosophical controversies among scholars (hidden beings), client-oriented purists, with strict rules, the legends and tales of most peoples, if not all, included references to creatures and races of birds, animals and non-traditional, and were extremely rare, and almost all peoples had creatures (their own hidden) of their own.

Diamond-like man / ape-like!

Puzzle G: brutal mother Abkhazia!
Is a creature belonging to the half-human faction, half-human
If we go to the East, penetrating the mighty Asian continent with its vast forests, its territories, seasonal vast expanses, so that we have in relation to the Turkish / Mongolian world, we will find a link to a strange creature that they call the creature of diamonds (Almas / Almas / Albıs), which is a creature that belongs to the faction (half-human), and among the list (human-beast), the platoon fairy is dubious in its presence strongly, which also includes the mechanics of my and big foot Bigfoot, etc. the diamond and the creature are relatively small in size, closer in quality to humans, it should have spread in Central and Western Asia from Mongolia, and even in the Caucasus.

According to the recorded testimonies, to prevent the challenge, they met a creature with diamonds, average in physical qualities between a man and a monkey, high, standing hair covered his body, except for the face and hands, and in most cases vegetable was mixed with it, and if he could eat anything, then when the fish got food.

Puzzle G: brutal mother Abkhazia!
Diamond creature, half-human, enigmatic and enigmatic

The first personal testimonies of the vision of Almaz's creature came from the Bavarian Johann Inter Johann (Hans) Schiltberger, who served in the army of the Hungarian king "Cossack" and fought with him against the Ottomans in the battle of Nikopol-Nikopol, which took place in 1396, but he was wounded and was taken prisoner by the Turks, who turned to them in parts of Central Asia. it is familiar to the inhabitants of the Pamirs, the Caucasus and the Altai Mountains, and even the fact that there are trade relations between local residents and objects of diamonds, in which they live like a loan. From the jungle, according to eyewitnesses, this creature does not have a registration fee confirming its presence in the books of a medical solution documenting its qualities, moreover, other creatures are quite real and still roam these edges, so why do you think that everyone the rest of the creatures are in fact while the diamond alone is some kind of mythological creature ?!

Several people took part in the consultations separately, on various occasions, including with the participation of "Don Myers" born in 1937, Russian "Ivan Ilyich", which took place in the Altai Mountains in 1963.

Puzzle G: brutal mother Abkhazia!
The diamond as shown in the brief description of the nineteenth century medical cluster
The more the views of the raven, which was witnessed by the great number of soldiers of the entire army!
This happened in 1925, when the Russian commander Mikhail Stefanovich Topilsky drowned his soldiers, pursuing the remnants of the anti-Bolshevik revolution, and who, having taken their lair from the Pamir mountains, where they were attacked by creatures similar to people and borderline at the same time, in one of the caves, which They dug up debris in it, found the body of a creature described by the author to himself, saying that whatever creature it was, it was covered with wool, although it was not a monkey, there are no monkeys in this area, and that it was a man. a man 5 and a half feet long, he appeared in the form of an elderly growth of white hair on his body, has the ability to measure and inspect bodies and make sure that this creature is not human, but at the same time, Time does not belong to any known species of monkeys or semi-monkeys!

(Note: the story is widespread on the Internet, the content is English and very loyal, and all the mobile ones are quoted literally from one source, so you look at some information that is not documented, then a search by the name of the Russian commander is mentioned, in English and Russian, I I do not receive information confirming this story, the basis of which, the closest person to Mikhail Stefanovich Topilsky described in this story, there is nothing in her story, something that belongs to the campaign in Alabama, characters, another world, biological who died in 1903 before a history approved for more than twenty years!).
Puzzle G: brutal mother Abkhazia!
Figure imagine to kill a mysterious creature diamonds in the hands of men of this theme
It has been reiterated that these views varied, and were often Berrachal or remained European, although all of these views combine all one recipe, which I did not write down on an image or one document to confirm that they did happen!

The average man's wife is rich: or as-as a person!

Puzzle G: brutal mother Abkhazia!
G mother of a wild tribe .. image!
Months of plot objects, Diamond is the story of Zana Zana, a female from the creations of
the diamond trading village of Salo, located on the banks of the Mukai River, Myk,
Abkhazia, a story captured by locals in the area of another Ochamchire on the east coast
of the Black Sea, and she was completely wild, and my leg, arm, muscular, muscular and
strong body are covered with hair, and larger than the size of a human normal, great
ability to withstand the cold, she brought her to a hunter to one of the smoking objects
called Die GAS, which, according to legend, kept her in his concern for three years, where
they stayed in a corral surrounded by a hedge of trees and threw food at her, even
subsided from the cruelty of the seal that fell upon them. And he began to master some
simple skills, such as grinding grain and money, since they became addicted to drinking
wine, grapevine, and then left him some freedom, which ended with the fact that an
unknown person gave birth to children in the village, maybe he was the owner of only
"gas", or else her children were brought out together with others, but of course she could
not take care of them, causing a number of innate instincts, in the immersion of the little
ones immediately after their birthday in the river as a result of the death of four of them ,
and of the children of "G" there are only eight, only four, two sons and two daughters,
another son and a son, who was told that she was the daughter of the owner of "g",
"Hoy" and "lightness", the gate ", which raised him as stepmother, I had to, but they did
not bore a surname, or another lost his in someone else's house.

Puzzle G: mother of cruel Abkhazia!
Porn: I looked for G and found her son!
Kids g are prone to dark skin and an unusual strong body, with a sharp, sharp nervous
character, but I stayed g, which is called exaggerating the problem, for the rest of my life I
was able to develop skills before college, unlike my children, and she died according to the
testimony of village elders in 1890. I stayed to mention the story of how "g" lived, flowed
past the village of elderly people, and made this information myself, as described the
funeral that was less expected when she died at the end of the 19th century, and all this
exciting information was told by a Russian researcher "Borisov porn" Boris Porshnev, who
was a historian specializing in social sciences interested in human origins.

This information rekindled the Hamas world of councils who tried to find the grave of "g"
with the help of some villagers who knew the location of the grave, but they were
replaced by the grave of the mother, they found the grave of her son named "Hoi" Khwit
", who died in 1954 and is in seventy works.

Helped to unearth the remains of Ji's son to prove that some of the ranks are scattered
throughout the rest of the ranks that live in that region, but this is still very far from the
described qualities of an animal or half-human, or, as the research team accurately said
(this is the skull of a semi-human and half unknown type)! ..
Puzzle G: brutal mother Abkhazia!
Hoi my son d: do not show the remnants of any properties of a non-human Meet the
"porn" of the grandchildren of D, where they notice the extraordinary power that they

use, so that the grandson is called "Areya" Shalikul, the strength of his jaws, so that he can hold his place for a person sitting On him !
The letter "G" was adopted by one of the Bigfoot creatures, or you are mine, the legendary one, who expected her stories so much in such zones of Central and Western Asia that some thought it might be a rare living version of the Neanderthal, the extinct Neanderthal.

Others, however, see that it has gone beyond what may be available, and the descendants of the descendants of the "g" of the dark may carry another explanation, the earliest and most systematic, of the history of the ghouls "G", and therefore they proposed an alternative that seems more persuasive and closer to research.

Isn't Ji the only African woman ?!

Puzzle G: mother of cruel Abkhazia!
The offspring of the others, her granddaughter, are these the African features inherited from her grandmother ?!
The story of "Gee" attracted considerable attention over a long period of time, and the professor of "porn" was another of the first elements of those mysterious stories to study them and try to find an acceptable and rational interpretation, although this was contrary to the beliefs of climatology, no, he was to "Gee" always as to one of the semi-human beings, since the woman of wild sex in him is more from the qualities of the animal and the wild than from the qualities of the human and humane, but is all this really so ?!

Could being "G" not / half human ?!

The answer filled the mind of the renowned "Brian Sykes" professor Brian Sykes, professor of human genetics at Oxford and World Report in his field, and the owner of a book set in a famous and multiplayer mode like his book (Seven Daughters and Eve) 2002, (The Curse of Adam : a future without men), 2003 and the book (bloody islands), 2006 and finally the author means that in our topic Today there is a book (Nature of the Beast) that was released in 2015

Puzzle G: mother of cruel Abkhazia!
Brian Sykes: Obsessed with the G story, find the birth of this puzzle in the sea
As an author of a controversial, as it is clearly encyclopedically perfect in his specialty, to which he devoted his whole life, Sykes decided to a legend (ogre d) worthy of his attention, to use his experience in the DNA of the remnants of the family of the Russian monarchy (the family of Tsar Nikolai Romanov, other Bolsheviks), he was a prestigious professor in clarifying this new riddle, and after tracing the rest of the offspring (Bigfoot) in parts of Asia, Sykes from Steph, who out of her safety said: take samples of their saliva to improve it, as well as the corresponding teeth of one of the sons of the deceased, and thus he got the impressive result - that "g" belong to West Africa.

No, it cannot be a brunette African woman, just a woman president, and maybe she was captured or kidnapped, by any means, brought to these remote areas, and she was probably a victim of a kind of secret slave trade, illegal, which was very bad for discussing a list of homework ... but one minute! ..

The wanted "Brian Sykes" I did not solve the riddle as we imagine it, but rather increased its complexity, in "D" Yes, 100% West African, but at the same time not owned by human authority to survive those ways now !!

In other words, the legendary woman is not from the security of the human Homo Sabine, which belongs to all human erect contemporaries, she is from the Security of another (funny) or (rare) or (unknown) !!

Puzzle G: mother of cruel Abkhazia!
The search results for aliens are related to health: could you be an "G" object that is not human in the first place ?!
So why is the paint deciphered after the riddle "g" still stands, looking for someone who does not offer a final solution, backed up by irrefutable evidence or is not surrounded by cops, mystery and fading!

Where, then, is "g":

If the breed is the population of Western and Central Asia, then how do you choose the recipes, recipes for its offspring, genetic, as well as from other people and races that still live there ?!

And if she really is, and according to the research of Professor Sykes, she came out of West Africa, then what about Jane from there, and to which security or human group to which she belongs, maybe "G" a woman from the people of the little unknown, genetically not different from the rest of the African population?

Puzzle G: brutal mother Abkhazia!
Sarah "Sartre" Bartman: can you be a G victim of different physical qualities like this?
We remember the tragedy of another, locked in his capacity as a person not often met, this is the story of a woman in the African poor "Sartre my Bartman" Saarji Baartma, who was subjected to the worst types of treatment and, due to the difference in the qualities of her body, Legislative and enjoyer are certain products in her tribe (Hoi), and this is known to the state of South Africa now, and how do you know that the booty was exhibited in the circus as a stranger, because they stretch your ass more than you do!

Likewise, if the "G" is not really common to mortal races, the end of the human chain is known exactly where it came from, so you may be one of the remnants of demi-humans who are supposed to be traditional climatologists who have gone extinct forever. ?!
The mystery of the letter "G" is usually not the most exciting thought about the mystery of the icon of St. George when he tries to portray a dragon - I mean specifically the dinosaur Baryonyx!

But this is another story, as he told the Godfather of his spiritual and deceased, great father.

Hollow Earth

Honey, let me think aloud about the fact that there is life beyond our dear planet Earth.

**And why all this money that the state spends on the actions of the West, led by the United States, in the task of discovering and searching for life outside the Earth? ... And where does this obsession with trying to discover life beyond planet Earth come from? ... Or are you thinking about the discovery of an alternative planet to our planet reckless in resources and the disclosure of mineral wealth outside the Earth? ..

Are all these missions sent into space precisely to discover resources or to divert the world from other life that lives with us on the same planet?

Yes, there is one more life on OUR PLANET, is it not the life of animals and birds, but what is inside the Earth, is it really similar to our lesson that these are rocks and volcanoes together, and hell underground, or is it the forests have collapsed and the thickness of the air and lives in the world of a hollow Earth ?!

Earthen depression
Researcher of the American Richard Byrd I learned from the life on a saw the proverb There is no smoke without fire, there is no fairy tale, no runes as a call without the presence of something true in these legends or myths. There are accounts of interviews received confirming the existence of life within the hollow Earth, including the story of Admiral Richard Byrd, who was the commander of the United States, was assigned to the task of exploring the Arctic, in the course of his search he discovered a large hole in the earth and was able to penetrate her with the help of the squadron, and he found himself flying high over cities, forests and forms of various animals, but he hurried again through the same hole into our world
... He has the same mission commander to make his escape to the South Pole and discovered another open in the South You in the world of the Hollow Earth.

Puzzled by this, I did not find a single media outlet where Westerners talked about these trips, the memories of the captain after the death of his story about his discoveries were published. But if what happened was real, why don't we watch movies - west of the talk or talk about the hollow Earth world or any sci-fi movie from this world, while most landmarks to be in space open up planets and stars.

And we, as a carriage, do not finance such excursions, and no program is alien, strong for discovering holes in the Arctic and in the south, so the West gives us what it wants us to know, it just keeps a lot of sketches of modern cultures hidden for us with you about nothing.

Returning to the Hollow Earth, which some of the theories emphasize the existence of a vacuum in a hollow Earth, the guide to these theories is that after the flood of our Lord Noah, the planet was filled with water on the other and climbed to the highest peaks of the mountain, the summit of Everest is in the earth, then When the flood ended, the groundwater smiled, but the question is where did all these quantities of water come from on the ground, which means that the presence of large cavities under the surface of the Earth can absorb this water.

Then where do the people, Georg and Magog, live, two of the many people, not the first and not the last, and many of them will not die until they see millions of their descendants, where is this incredible leap of them? Where to live on our land? ..

Far from the jurisprudence of scholars and theories, we find a verse in the Qur'an that emphasizes that the Earth is the land on which we live, the ascended one says,

"God, who created the seven heavens and out of the earth as they are, deals with him between them to know that God is capable of doing all things and that Allah embraces all things in knowledge."

This verse confirms the existence of seven heavens above us and from the Earth, any seven see you two

And let's ask ourselves, where are the seven lands two? ... And the answer most likely lies in the fact that it agrees with the theory of a hollow Earth, that is, that our planet on which we live, where there are cavities and other worlds.

Earthen depression
Layers and the return of the other under our known world There is also something very strange, and the published reports of the existence of cooperation were established between the inhabitants of the hollow Earth world and the German Nazi government during World War II, this was one of the reasons for the thinking of the military Germans during that period. , so this collaboration or communication with the inhabitants of the world of the hollow Earth provided the Germans with access to advanced science, especially since we know that the Germans were the first to connect to guided missile technology of nuclear and other ahead of their time, this confirms that the inhabitants of the dungeon are at the forefront of science and technology.
With the help of these guides, let's ask the question of what is happening to the inhabitants of the hollow Earth now, Wei, do you see the United States becoming an alternative to Germany in this cooperation? Is what they say about aliens and flying saucers part of the secrets and techniques of this cooperation or an attempt to show off, especially since more and more observations are taking place in the United States? ... Allah Almighty and Almighty.

War of the worlds in the Nuremberg sky

The phenomenon of flying saucers, unidentified and the history of their engagement is not new, but as old as history, where there are ancient scrolls and photographs on the walls of temples, dating back to ancient civilizations, like the Sumerian pharaohs describe feelings similar to what we today call ufological identity (unknown flying object). Despite the ambiguity and uncertainty that surround this phenomenon and exist among experts about the possibility of its existence or not, only the fact that the number of views on UFOs and the retention of aliens at different stages is increasing is very large.

War of the worlds in the skies of Nuremberg

UFO sightings and aliens have long been controversial since ancient times
Accidents and malfunctions of aircraft dishes can be classified by degrees, or types, there
are, as it were, close collisions of the first kind, second, third, and often can reach the
highest degree of it or what is called type IV and V (close collisions of the fourth and fifth
kind) ... Experts and those interested in studying plane anonymous bodybuilding accept
these classifications based on the degree of confrontation or encounter, possibly the
severity that results from them as well. A large number of aircraft opinions and objects
and respect for aliens are authentic in the modern era, ranging from individual
consultations or groups that arrived in dozens of them, to the maximum of its distinct
type IV and V, both of which involve two cases of abduction or direct contact , and she
could not get the governments of the world of especially large powers for her silence and
her about everything about this phenomenon until not so long ago, it was only in the last
few years, starting in 2007, the experienced phenomenon is a kind of openness to the
limited around governments large countries especially in relation to the environment on
my planet. will have to own and release files.

War of the worlds in the skies of Nuremberg
Some graphics for space bodybuilders around the world
The French government initiated in 2007 the publication of a "fact-finding king" on the
problems of flying objects that existed during the time of the French space agency. The
British government's position was to disclose and make public its dossiers on the
phenomenon in 2017, where it also published documents pertaining to the government-
funded investigation and intelligence gathering department of private investigations of
unidentified flying objects under a program that lasted for about 50 years, starting in
1947. , this is an interesting addition to the main work of collecting information about
flying objects. The United States of America also began gradually releasing a volume of
aircraft dossiers at the end of 2017, and has approved the US Department of Defense
(Pentagon) for a secret special study program.

The narratives and types of recorded objects on the aircraft vary in the degree of bizarre
mystery surrounding the aircraft, but most of them share many common points, especially
with regard to the nature of the incident and the temporal and spatial environment, in
addition to having great similarities in the details of the incidents.

War of the worlds in the skies of Nuremberg
Drawing on a tree never happened in the sky over Nuremberg
The story of what happened in the sky over the German city of Nuremberg in Bavaria does
not look like the numerous recorded incidents, but it is more than that and, perhaps, the
one and only of its kind between accidents and reliable.

At dawn on April 14, 1561, he awakened the inhabitants of the German city of Nuremberg
with strange sounds and a thunderous explosion of rage. The turn of the people's anger
towards the sky, transmitted to the network, expressions of horror and anxiety clearly
appeared on the faces of the city's residents after observing hundreds of objects: the
planes flew at high speed in the atmosphere of the city, some moved in the form of
squadrons and single ones. Flying objects have various shapes and designs, some
spherical, some cylindrical and cruciform, with sizes varying according to eyewitness

accounts. Flying objects launching blood-colored rays, the presence of smoke rising from some aspects, and the roar of explosions as a result of the last number falling and hitting the ground by them also appeared in the description, there is also a description of the use of two large sickle-shaped bodies and another huge triangular one.

The first description of the incident appeared in a report by German journalist Hans Klass (The Glaser) in the air, where he printed the details of the incident on a woodcut (woodcut) panel of his own making, as documented in eyewitness testimony and other details about the incident, as the local press in The city (Nuremberg Gazette) documented this incident based on the Hans-class report on the testimony of the city's residents. All documents related to this case, including the waves class paintings and present in the Central Library of Zurich, Switzerland.

War of the worlds in the skies of Nuremberg
Hans Klass manuscript
What is stated in the report of Hans and his description of the details of the situation and suggests that what happened in the sky over the city of Nuremberg was a kind of conflict between the two forces, may belong to other worlds. If we take into account that people in the sixth century out of ten did not have the usual concepts and aviation technology, then we believe that the description of people set out in the report of the journalist Hans-class, in terms of the shapes of objects, approached the description of aircraft and aircraft designs recorded in the modern era.

The recorded details of what happened in Nuremberg are considered one of the best documentary evidence of the phenomenon of flying objects, having adopted many experts and those interested in the Acuvue field. ...

The accident in Nuremberg cannot be classified as normal from a series of incidents or sightings of flying objects, they are not her private life, being knocked out in half a minute from flying objects in large numbers in case of war and conflict between them, they are probably the first a written description of air combat between flying objects in modern history.

Grimlins: mysterious creatures that sow terror in the sky

There is no doubt that the archives of the history of aviation and the Air Force around the world are rich in events and strange stories that do not eliminate mystery and intrigue at the same time, especially those that tell of views and mysterious equations that are difficult to interpret based on known physical laws.

This is the story and, perhaps, the most mysterious of those that took place in detail during the two world wars, which were documented through messages registered by the aviation authorities and the air forces of the belligerent states, and the propaganda of the appearance of a strange object of aggression, activating the terror between pilots and aircrews. the mysterious object was shot under the name The Crime of Two Gremlins or Green Gremlins in the plural.

Girls: mysterious objects spread terror in the sky
The mysterious object is relatively small In the midst of air battles and bloody collisions with enemy aircraft, where sounds of explosions are heard and actions filled with smoke, pilots and crews of the aircraft themselves find themselves in the face of a different kind

of enemy, a creature or object of a mysterious at first relatively strange shape, an aggressive Berserker, appearing suddenly, both alone and in large numbers inside and outside the plane, talking at high speed, attacking and all that, destructive and chaotic in all parts of the plane.
The multitude of stories and messages based on the communication of non-moving pilots registered with the Centers for Disease Control indicate that there is a noticeable connection between the occurrence of technical problems and, possibly, the fall of a large number of aircraft with the appearance of this object or objects with similar technical characteristics. significant and is a history of a registered incompetent third party or just one country, but involving aviation authorities around the world.

Interesting stories about these creatures lead to a story with a signed detail: in 1923, the birds of the British survived after his plane crashed into the sea, as described in his report of the crash incident, he was attacked by strange small creatures that suddenly appeared inside his plane and caused damage and destruction to a number of existing equipment inside the aircraft, resulting in engine failure and aircraft crash.

Girls: mysterious objects spread terror in the sky
Attacked by strange small-sized creatures suddenly appeared
American pilots had their own stories of these alien creatures and the history of some of these sightings and incidents dating back to the thirties of the last century.

One of the American pilots knows his name, lab L. W, who was one of the roaring heavy B-17s in World War II, sees a terrifying face with these creatures that appeared in various parts of the plane after takeoff a short time later, where he describes, that he first heard strange sounds inside the engine and equipment inside the cockpit, and then was surprised by the existence of a terrifying creature seen from one of the side windows of the cockpit, and he also says in his report on the situation that he was shocked and confused by the harsh of the two. when he saw one of the creatures sitting on board the front of the plane and another one quickly move to the front window of the plane and use it to try to crush them.

Girls: mysterious objects spread terror in the sky
The creatures are short, with red eyes and two pointed ends. In his report on the shape and technical characteristics of objects, the pilot indicates that she is small in size with a length of about 3 feet or less, her eyes are red and radiant, her arms are unnaturally long, with the skin is of a gray level, the mouth is protruding, carefully pointed and long pointed.
The dangerous incident, recorded by the US aviation authorities, is the attack on the pilots and crew of a transport aircraft of a US military blender of 13 people according to the description of the object, similar to the series returning the bus in the late summer of 1939, a few days before the joint birth of life united in the world war, where the watchtowers at the airfield at the San Diego base of the US Navy project branch in California need to be urgently removed from the plane after about three hours of takeoff, then the plane appeared in the sky of Cairo after breaking contact with it for several hours, and it was destroyed .tried to make a hard landing, and traces of casualties and destruction remained on the outer hull of the aircraft. The report recounts a gruesome

scene in which his rescue crew was seen and a recording of the crew as they entered the plane after a safe landing, where they found the flight crew's bodies ripped apart and arousing fears like predators, the co-pilot who could land the plane, he was the only one alive, but he soon died in the hospital from his wounds eloquently.

Despite the fact that most reports from pilots are characterized as aggressive and this creature is not, this is exactly what the pilot and inventor American Charles Lindbergh told Charles Lindbergh about the incident with him, which he published in his famous book The spirit of St. Louis in 1953 refers to to a dissenting opinion, claiming that the creatures that appeared after the nine-hour flight, and she helps him on the tier constantly talk to him, and he showed you information specific to air navigation during the long-haul flight across the Atlantic, which he did in me.

Describes Lindbergh those objects that are foggy on her or on the body of the scooter are slowly replenishing, very cleverly, which can give the impression that she may be ghosts in nature or objects from another dimension - one

It is worth noting that the recorded observations were not limited to the pilots and crews of aircraft of a particular state, but included various states, for example, Germany and Japan, however, a large proportion of reports of problems with attacks by these creatures that arose before the aviation authorities were from British pilots who entered World War II.

Girls: mysterious objects spread terror in the sky
Joint delivery of items of mysterious Old English folklore.
The label or term "Green Gremlins" spread significantly among British pilots in the twenties of the last century in reference to those creatures that hid small objects to some extent and pointed ears and large sizes, and has since become part of private aviation folklore, but the original delivery. apparently has no ancient roots, can be derived from the English word an old crime of Greme, which describe a certain kind of creatures from the world of sprites and demons.

From what was published about the appearance of this object on the plane, follows a report published by the British newspaper The spectator at the beginning of the twentieth century, where, according to observations and incidents relating to this mysterious creature and the longest registered property between 1917-1918.

In April 1942, the magazine published an article in the RAF magazine, entitled matter series by Hubert Griffith The Gremlin Question. Hubert Griffith is dated by accounts of the presence and appearance of these mysterious objects.

Girls: mysterious objects spread terror in the sky
Recognition of the existence of this phenomenon, but without explanation of the reasons does not occur
In 1940, the British Air Department of the existence of this phenomenon formed a special committee to investigate and establish facts about the problem and, unfortunately, did not publish anything about the results of this research and research, but the Ministry later

issued an oversight of the pilot officer Percy Brunei Percy Pruhn with dignity helped the pilots on how to deal with and get rid of these creatures when they appeared, entered the pulpit and explained the species and forms.

There are many opinions about the existence of these creatures, apparently, one of the pilots during the Second War judged the Vatican over what happened to the pilots. Associated with psychological pressure is anxiety about severe chronic exposure to pilots during the war and has led to cases of hallucinations of states of hallucinations, but the presence of some facts can make this interaction weak, such as the presence of a large number of reports of pilots experienced high-level professionals in air war with no previous records refers to the presence of the same situations and has a tradition similar issued by the pilots were in Rahal air Steiner military

Girls: mysterious objects spread terror in the sky
Movie quote personal episode in famous 1984 movie
Seeing the presence of green care in the science of hidden animals, cryptozoologists tend to confirm the existence of these creatures and see them as a kind of naval object called the Goblin goblin, while others think they are aliens, they associate their visit with the bodybuilding planet and anonymous.

Despite the contrast between the reviews about the existence of this hidden creature and the big mystery that surrounds the subject, and this, in addition to trying to find an explanation for the trend, it is logical that what actually happened to the pilots and aircrews, but a large number of reports and observations by pilots and crews planes made from this object are part of the heritage and history of world aviation .. as well as film history, as these little creatures are terrifying - and hilarious too! - In several films, the film achieved widespread recognition and became one of the icons of horror cinema.

Ingo Swan: the man who explored Jupiter

In the midst of the Cold War, competition between the United States and the Soviet Union intensified in many areas, as the invasion of space and the development of nuclear weapons and other spheres of activity as a parasite are controversial, and that one of its branches is the important thing known as "found after" any ability that can mediate the description of a specific geographic location by transmitting consciousness to the destination.

And that was exactly what he enjoyed with the strange man "Ingo Swan", born September 13, 1933 in Colorado with a painted heart. He was a psychiatrist and all, and he focused on the parasailing case besides Russell Tyrgu and Harold Beethoven.

Ingo Swann: The Man Who Explores the Depths of the Buyer
Ingo Swann
In 1973, it was founded by the Academy of Sciences and Media in the case of a keen runner for any research concerning the world, based solely on the involvement of Stanford Research Institute in this area is a culture shock both scientific and scientific level. the buyer of experiments designed to test cognitive and mental strength for the long term, and were subjected to severe ridicule by scientists on issues of mutual interest, interpreted by mainstream scientific principles.

Trade

Ingo Swann: the person who explores the depth of the customer
You trade on the basis of perception after
Established trade, which lasted for 20 minutes at the Stanford Research Institute in collaboration with the CIA, where he was a broker, Ingo Swann had to present through the ability to perceive simply curious scientific data from nowhere and unknown, and the choice fell on the buyer's planet to be confident in the future.

On the evening of April 17, 1973, Dr. Russell Tirgu and Harold Beethoven heard the swan side to observe the planet Jupiter and its satellites before NASA's 1979 Fusion 1 probe visited the planet. According to Sue, his ability to see the customer took about three and a half minutes.

The meeting cited many of the attributes of Jupiter, describing its atmosphere as a very huge hydrogel coating, as well as the choice of the trajectory of the approaching probe, which is expected to be from 80,000 to 120,000 thousand kilometers from the planet's surface, the swan claimed that he sees crystalline puddles in its atmosphere that are almost clouds. perhaps, like the rings of Saturn, the Viola 1 probe later confirmed the presence of buyer episodes, despite the fact that these episodes are not in the planet's atmosphere. However, the statement about the presence of crystals present in the atmosphere is supported by observations from the NASA Galileo spacecraft, which revealed the components of resistance of ammonia ice crystals in the northwest corner of the Great Red Spot to the buyer.

Below is a copy of Swan's sayings in 1995, after 22 years of making art

Ingo Swann: The Man Who Explores the Depths of the Buyer
It has not yet been possible to identify the Swan for obtaining information about Jupiter!
6:06:20 "There are crystals very high in the atmosphere ... they sparkle. Maybe in the form of a cloud of crystals, maybe like the rings of Saturn, although not as far away as that. Very close in the atmosphere." (The phrase is clear.) - I bet it will reflect radio signals.

6:08:00 "Now I'm going to go through these clouds. It's a very good feeling (laughs). I've said that before, right? Inside these layers, the cloud, these layers of crystal look very beautiful outwardly. In the interior they look like rolling gas the clouds are the yellow light of a strange rainbow. "

6:10:20 "I have this impression, although I don't see that he is liquid."

6:10:55 "and then I came across a cloud cover. The surface looks like dunes. It is made of very large crystals, so it glides. There are also huge sports like Earth winds prevailing, but very close to the buyer's surface. and it is this product that the horizon appears orange or pink, but click on it to look yellowish green.

6: 12: 35 "if you look to the right, the mountain range there is huge."

6:14:45 "I feel like there is liquid somewhere. These mountains are so huge, but they are still not finished through the cloud cover crystal. You know I have a dream like this thing where the cloud cover was big the sweep of the whole sky that the grain sand is too large with a color orange, it has a polished surface and is similar to the color of amber or like obsidian volcanic, but it is yellow and not heavy, the wind blows up. Slip. "

6:16:37 "If you turn around, it's all very flat to you. I mean, if I felt that if a person stood on the sand, I think it would change in them (laughs). This feeling comes from the sensation of fluid.

This was established from data provided by Inga Swan after research by the NASA space agency with 1973 Pioneer 1 (after a few months of experiment) 1974 Pioneer 2 and Explosion 1 and Explosion 2, it turned out that there was an obvious coincidence between the information provided by Swan and research by modern science, including:

Ingo Swann: The Man Who Explores the Depths of the Buyer
The newspaper's image related to the Swan experiment
1 - buyer's loop discovered in 1979
Telling Time magazine 1979 March 19: you know the probe to the radiation Dust coming from the depths of the clouds and the most amazing part is to detect the presence of a thin ring flat planet was surrounded by great, says Red Smith, an astronomer from the University of Arizona: here we stand open our mouths, don't behind in Breaking overself.

2-liquid detection on the buyer's surface

The scientific journal News 1976 July 17 said in this regard: look at the earth's crust, so that the buyer of gas, mainly hydrogen, and during high pressure turns hydrogen into liquid at a depth of 25,000 thousand kilometers.

3 - Discovery and presence of a mountain range

The words of Swan, whose mountain range led some academics to ridicule the experiment, was to stare at the buyer that he is the nature of gas, but scientific discoveries later proved the sincerity of his speech, where he told Science News on July 10: It looks like That the Great Red Spot (one of the most famous atmosphere landmarks for the shopper) is stuck on top of a mountain a wonderful climb.

4 - The presence of spectra of colors in the atmosphere, which are selected only precisely marvelous

She told Science News on March 10, 1979: "Who makes the shopper's convolutions more amazing is the spectrum of red, orange, yellow, brown and even blue.

And there are many sources, journals and scientific journals that published the scientific facts that Swann mentioned prior to several years of discovery.

Gate of Time: Santiago Flight 513 Incident

The story of the crash of Santiago Flight 513 is a true story, possible and mysterious too, where he describes the sudden appearance of an airplane after nearly 35 years of its disappearance! ..

The events of the history of the city of Aachen in West Germany began earlier, the plane took off from the city airport on a regular trip on September 4, 1954 carrying 92 people, including a crew of four heading towards the city of Santiago on the other side of the Atlantic.

Having turned off all communications on the plane a few hours after takeoff somewhere over the Atlantic Ocean, the plane disappeared without a trace and heard nothing more.

Appearing out of nowhere.

Gate of Time: Incident with Plane 513 Santiago
The plane took off in normal mode from the airport of Aachen
Strikingly working in the control tower and witnesses of the return flight of the aircraft on the day of October 12, 1989, and suddenly from nowhere above the airport of Port Allegro Port Alegre in Brazil, and also landed safely on the ground of the airport after the process of the rotation of the sky over the airport.

The control authorities at the airport did not use any advice from the flight crew during their sudden appearance and fall, which only made it more mysterious and exciting to complain and have the airport authorities rushed to send a group of people wanting to inspect the plane. The members of the security team who boarded the plane were struck with a harsh horror as they inked their eyes on the area of horror and horror inside the plane, the scenery of 92 skeletons of people sitting in their seats on the plane, including the skeleton of the aircraft commander, Captain Miguel Victor Kirk, Miguel Victor Curie sitting in the cockpit, clutching the seat of the aircraft.

One of the first to write about the incident is a report published by the Weekly World News tabloid magazine in its November 14, 1989 issue by journalist Irwin Fisher Irwin Fisher.

Gate of Time: Incident with Plane 513 Santiago
Passengers and Crew Skeletons The crash of Santiago Flight 513 sparked controversy and debate among stakeholders who hesitated between locked and your limits and ensured that the discussion also tried to explain what happened with the adoption of various theories.
It was difficult to explain this situation. he was faced with difficulties and ambiguities regarding the details of the incident, exciting many questions, for example, how to explain the sudden appearance of the plane from nowhere and the safe landing of the plane with its skeletal crew.

Researchers who have tried to find an explanation for this conversation are Dr. Celso Atelier Dr. Celso Atello is a supernatural researcher who stated that the plane must have entered Dodge's gap, but he was never able to provide an explanation to achieve everything there was on the plane for the skeletons and how to enable the skeleton to land the plane safely.

The Brazilian government, for its part, formed a search committee and facts about what happened, but it did not release a single official statement about the progress and results of the search and investigative actions, while, as the Brazilian aviation authorities confirmed, the incident pointed to some aircraft details that suddenly appeared in the sky and landed safely.

Gate of Time: Incident with Plane 513 Santiago
What has already happened to the plane? Surviving the work of the committee on government research and results in the speed range of launch points and at the same time anger, where the student was much interested in academics, opened up the scope for the government for the participation of civilian researchers in the search. Physicist Professor Rodrigo de Mancha Rodrigo de Manja was one of the first applicants for the government to disclose information, which said that masking any information about the incident was a great injustice in relation to the right of science and humanity, because in the event of an aircraft entering the temporary Time wrap would change the way we look at today's science and the world.
Whether it was an accident, Santiago's 513 plane is a fact, or science fiction also supports skeptics, it is the presence of a scientific basis in the network that made them possible even if science is still in a physical system; many of the science facts have now been dealt with in previous science fiction.

The theory of bridges at time-space gates could give Apple the best and most powerful home to mitigate the disappearance and sudden appearance of people and vehicles in history.

Can you explain, Mr. Einstein, what happened?

I remained the general theory of relativity of Albert Einstein as a strange and mysterious between the theories of classical physics even after many years of its publication in 1915, perhaps due to make these theoretical ideas and concepts revolutionary and complex in physics, which once led to a radical change in the concept The universe that existed earlier in theoretical physics. Einstein predicted in his general theory of relativity the presence of curves, or regions in space and time, in other words, the interdependence between space and time, these predictions led to the creation of scientific concepts of a new club of black holes and wormholes.

The idea of introducing time as the fourth element into the equation to describe a spatial specific event led to the emergence of the concept of the fourth dimension, the fourth dimension.

Gate of Time: Incident with Plane 513 Santiago
The theory of relativity brought forward new concepts in physics, including wormholes
The idea of paths or bridges connecting different points in the concept of space-time was put forward by Einstein in 1935 in collaboration with the world physicist Nathan Rosen, Nathan Rosen in the Scientific common research work called Einstein Rosen bridges.

The Einstein-Rosen bridges along this corridor are default parameters, and they also suggest that these bridges can connect between points in the universe, reality, or be considered multiple, not much is known about the nature and composition of these holes and what may be contained inside , but there is a premise that suggests that it is unstable and the possibility of its rapid breakdown, this is in addition to their high signal content and the possibility of the presence of strange exotic matter.

Inside it, making it dangerous to travel through it.

Despite the survival of the idea of lanes in space-time within existing theories, they represent a closer knowledge of the concept of time travel - an incredibly different or parallel parallel universe, which makes it fertile material for a book about cinema and science fiction in the past and present, where dozens of famous sci-fi novels and films are being released that have been the subject of their journey through these transitions to supplies and various

Hydrocephalus: People Without Brains

Is there a need for a brain? You should be amazed at such a question and consider it illogical only what I have learned in curricula in schools and universities, which is the inevitability of life and damage to one of the parts corresponds to fire in activity, physical or mental.

But there are striking cases where the brain does not exist or is limited to a small denticle of nerve blocks, the most famous of such cases is الهيدروسيفالوسHydrocephalus some rare cases are called the pseudonym Anencephalus this term refers to the case of the loss of a large part of the brain of the biological brain so that the skull is almost empty, except for a layer of tissue perfect on the walls of the inner skull, people who are stylists in this case, you can stay alive and excited your life naturally !.

Headrests: people without brains
Anhidrosis is an accumulation of cerebrospinal fluid, which puts pressure on the brain and leads to light, both in the picture of the problem (ascites) and, to a lesser extent, the child was born, but the child has blood .. the upper part of his head is absent, and lives in general I have a more surprising discovery, where the brain is completely absent from the skull, so it surprised scientists to prove pieces that the activities of various psychic like memory, logic of thinking and intellect are not related to the brain!
The months of cases are the ones that appeared in the 1980 science Scientific journal, telling a story that hit the University of Sheffield, UK in 1979, where the income of one university student at the age of 26 before the doctor's office was cold. But the doctor noticed, that student B, who had various degrees in the award-winning terrace mathematics, was kind of normal, but the head size was larger than usual, so the doctor only had to send the student to the doctor John Lopez, who was the most important neurosurgeon at the hospital in Sheffield. destroying the nerves.

What kind of doctor Lopez was surprised that a student who increased his wit by about 126 IQ points, who showed no signs of mental illness after being examined using a computed tomography, the cat-scan showed that the Head opens into the brain!

Instead, it was full of fluid, except for a mass of medulloblastoma, a little skeptical about the site. The thickness of the canvas is a few millimeters, not a few sets in a natural state!

Dr. Lopez knows about the existence of such cases, when a change in the cluster of a stroke occurs just in the head, but whoever follows his interest in this case misses this young student in his scientific research with a high IQ of all this and a head stroke does not exceed 100 g in comparison with a natural brain weight of 1500 g, which makes you seriously think: is a brain needed for this ?? !!

Another case of a different kind

Headrests: people without brains
The ancients believed that the heart is the center of thought, conscience, and not mind ... the pharaohs' brain, finds it or heart, unscrewed the remains of a mummy and without them I do not worship the dead in the afterlife.
From the experience of life and all of us today, we collect on faith the scientific generally accepted that a brain so severely injured or traumatized as a result of oxygen penetration into the skull will affect the functionality of Coachella's legs or blindness and even

death. But there seem to be exceptions to what might seem like the former is not always true, as we will see in one of the most extraordinary cases that speak of human ability to succeed, resilience, and struggle to survive.

The Case of Phineas Gage Wonder

It was Gage at twenty-five years old, when, as the head of a workshop for the production of works in 1848, and one day while working in a workshop for breaking rock formations requiring blasting of deep powder, he was Gage of his hand and shoved gunpowder into one of the holes. drilled into the rock using an iron bar, when he called him his friend, from afar, and as soon as I turned towards the source of the explosion sound, the tongue was not perfect, but it was enough to make the penis fly out of his hand like a Rocket to hammer away.

This penis that was launched in glimpse and quickly pick the head of the GIA without who knows! What I got after a few minutes is that those gathered around the man who fell to the ground under the pressure of the explosion were shocked and stunned when they saw a large hole in his skull as he woke up from his sleep, behind the conversation with the participants and it was something that was not there !.

I chose the penis of his skull, a history of a puncture of a wide, medically damaged frontal lobe of his brain, his frontal lobe is almost complete, and that it still connects neurology and medicine in general, to this day, because Gage got up from the ground behind talking naturally , but he knew that he was severely injured due to the heavy flow of blood.

Headrests: people without brains
The case of Phineas Cage is one of those rare occasions where blood damage rises dramatically along with this living source of the rift player long after the incident ... and is shown here with a penis that chopped his head off.
During the care period for this latter, the great Gage noted that he gradually began to suffer from a case of fever, then achieved a lost ability to speak clearly, and then a case of non-reaction with the ocean, and then went into a dormant state for several days and remained in this state until the beginning October, where he gradually recovers to make a full recovery by the end of November.

To the surprise of the doctors who examined him after he had fully recovered, they found that he enjoyed all the elements of excellent health, with the exception of loss of vision in his left eye and soft patches on his head, where the skull is incomplete, a strange thing in this case. is that the hole was too wide to be seen immediately and clearly, and this is what made Gage work in the circus. most of his life he spent across the head and paws.

Matt Gage suffered from epilepsy after many years, believed to be the result of an incident early in his life. The skull, they are on display today at Harvard, are attached with the metal bar you choose to hang it.

His posture is still strange even today in the medical community, since this is a strong slap in the face for some, where there is no place for miracles in this world due to climate, everything has a scientific explanation. If so, then let them rejoice in this scientific matter!

Breathing: supernatural bodies that defy nature

Despite the scientific development that the modern age has seen, people are still unaware of many of the same phenomena of exceptional hacking that appeared on the hands of many millions of people in different eras, indicate the mysterious power that the human race possesses, and still burn houses its scientific systems of physical pressure, since the exclusion and classification "as pseudo" is the fate of situations and phenomena arising from the customary for the region imposed by the institution of the Academy of Sciences.

Among these abilities and the extraordinary is what you know (breatharianism) anyone can live without food or water for long periods that can last for years. He can get people to live in life by relying only on the air - and some practitioners believe that these people can only survive in the light of the sun! ..

These Dahers, especially among the Sufis and saints who maintained their health and remained for long periods of fasting from food and drink, reported many examples of them in a book entitled "The Autobiography of One of the Yogis of the Two" "Paramahansa Yogananda."

Development: bodies of nature's miraculous movement
Hindu guru Prahlada Jani's mystics thrill perplexed scholars
The most famous of these is the mystic of the Hindu guru Prahlada Jani Prahlada Jani, born on August 13, 1929, who arose in the village of Shraddha in Mass.

In accordance with the committees, he left his home in the sixth and went to live in the forest. At the age of 12, Jean's experience of "religiosity" made him follow, in particular, the worship of the Hindu gods Emba.

The anxiety of this old man, who has lived without food or drink since 1940, when he thinks that the gods are supplying him with a fluid stream through a hole in the sky, which allows him to live without food or water, has piqued the curiosity of many scientists around the world.

I had numerous medical tests on Jan under the supervision of "Dr. Sudhir Shah," a neurologist at Stirling hospitals in the heart of Ahmed Abad who has studied many people with extraordinary abilities.

I confirmed to researchers that Jean's ability to live a healthy life without food or water throughout the test, although he did not publish the results in a scientific journal, as usual.

And when it was a question from the spokesman of the scientific group in charge of details, it was announced that the study would be "secret" until the results were determined, and the text of the official press release by another group, which could not disclose any inside information, would only report at the discretion of the chief investigator.

2003 tests

In 2003, Dr. Sudhir Shah and other doctors monitored Jean's case at hospitals in Sterling for 10 days, when he was placed in a closed ward. Doctors say he never leaves the room with urine or feces during the observation period, but there appears to be a floor in the bladder.

A hospital spokesman said Jin was fine, but he indicated that the hole in the sky was abnormal.

2010 tests

Development: bodies of nature's miraculous movement
Was subjected to many tests
In 2010, Prahlada Jani's monitoring study was conducted by the Institute of Physiology and Medical Sciences of the Ministry of Defense (DIPAS) and the Research and Development Department of the Ministry of Defense (DRDO), where they put strict isolation and level control, video surveillance for 15 days. And pull it out for MRI, ultrasound, X-rays, and sun exposure under ongoing videotapes. Clinical, biochemical, radiological and other periodic checks were also carried out.

It was said that Jean's contact with any form of liquid only during the study period occurred during gargling and bathing in some cases, starting from the fifth day of the test. AND

Close the toilet in the room, Jean-data that does not need to urinate or defecate.

Fifteen days after the test, which was not directed at food, drink, or toilet, it was reported that all the medical tests performed on Jin were normal and described by researchers who were in better health. than anyone of his age.

The doctors said that despite the fluctuations in the amount of fluid in Jean's bladder, Jean must be "able to generate urine in the bladder," and with this he never urinated.

Based on the reported levels of leptin running, working, namely hormone-related appetite, suppose the DRDO researchers suggest that Jean may develop extreme adaptation to hunger and thirst.

In September 2010, Dr. Shah, who was invited by scientists from Austria and Germany to visit India to do more research on Jeanne, also shows scientists from the United States to join the search. Jean is interested in cooperation in further investigation ..

يرمدstated at the end of the testDAPIS: "that the results of the study can be of great benefit to humankind, as well as" soldiers and disaster victims and astronauts, "and all of them may be left without food or water for an extended period of time"

Having dealt with many documentaries, Jean's case also appeared in an interview on the Discovery channel called "The Boy with the Power of God", as well as on the independent television network ITN Info about the 2010 tests and in many other world media.

Other cases

Development: bodies of nature's miraculous movement
Theresa Neumann's attitude has many supernatural properties
Jean-the only one who is known for his ability to distribute food and water has not read, it is between those who sparked controversy in Europe of the twentieth century that there

is a Catholic monk Teresa Newman, who was born in 1898 in the village of Bavaria, Germany, in a large family with a small wealth.

This country girl amazed people with her ability to starve for years without putting anything in the stomach.

In July 1927, Newman's charges against her home were examined. Physically proven and physically tested physician Otto will support the four Franciscan sisters for fifteen days (July 14-28). And he didn't notice that Newman had eaten something.

Theresa Neumann is said to have eaten no food or drank any water from 1923 until her death in 1962, and many books about their supernatural ability to survive long periods without food or water.

There is also an Egyptian man named "the Ashmawi thing" who contained a story in the realization newspaper in 1882/4/1, this man did not eat or drink for twenty years! That he died but lived his normal life. Once the "Laroussea" thing of the Grand Mufti of Egypt, then the book Ashmawi's thing in an isolated compartment for two months without any food or drink. And considering the state of his health !!

Are there really ghosts in the mirrors? .- Yes, that's right, here's your guide!

There are legends and myths about the mirror, the most famous of all the Bloody Mary tales, about that crazy woman with shaggy hair that shows up on the page of the mirror when I called her three times and hold with my candle inside a dark cabinet, showing the knife in her hand , huge for the souls of those who dare to call her and them to the inside of the mirror to stay there forever ..

There is another myth that you are talking about the opportunity to see your spouse, and it increases the likelihood that if you sit in the room for a few extra seconds focused on the center of your face for a certain period of time, you will inevitably see your partner appear behind. you, will be angry with you for bringing him in, and can hurt you a lot.

There are ghosts in the mirrors, right? Yes, that's right, here's your guide!
Bloody Mary .. more famous and panicking ghosts in the mirror
Some say that this myth is unfounded, while others see that a mirror is not just glass in which we see a reflection of our image, but it is also a gate from the floor of our world to another world, filling it with mystery and secrets, these gates can be realized for one reason or another, they had to take possession of his objects, terrible hearts filled their

eyes with horror, and that is why he advised elders after that to look in the mirror, because the longer a person looked, the more chances to cross these objects, special for our world, cause all kinds of misfortune, pain and problems.

To tell the truth, believers that mirrors do not give rest, they are, perhaps, closer to the right, here I am talking not from the point of view of propaganda of differences, but from the point of view of purely scientific logic ..

But hey, no knowledge "and" that's who believes in the existence of Bloody Mary !! ..

This is optics, which is one of the oldest sciences, which in the means comes out of the visual, which can be defined as seeing the risk arising from the failure of the brain in the analysis of data, transmitting it to its gaze properly, a function that you just see, since an apple is a function of the brain, it can deceive the brain and keep it easy, the simplest example of the image below, where there are 12 points of black color, I dare you, dear reader, to see and there's no time, because whenever I chased one of them with my eyes , I chose another one as a talisman! .. Facebook Facebook sparkled this image with a sensation when he posted the world of Japanese psychology cache Kitaoka on his Facebook page for the first time, just like I took a photo of a dress that was posted by a girl some time ago on Facebook and spicy world at battle for his true color !.

There are ghosts in the mirrors, right? Yes, that's right, here's your guide!
Image of the Japanese world .. the more you look to select the rest of the points
Opto is trying to explain the optical illusion in a scientific way, and he must have seen the appearance of monsters and terrible objects in mirrors, this is just a trick visually, no more, but tries to interpret them on the basis of an optical illusion called fleeting or peripheral fading (Peripheral fading) and also known as a faded rose (Troxlers fading), referring to the name of the world, discovered in 1804, which states that when looking at a set of image points or organizations with an emphasis on a particular stimulus, such as a pointer or attitude, over a short period of time, the surrounding points will gradually fade in and out.

There are ghosts in the mirrors, right? Yes, that's right, here's your guide!
An example of a vanishing 3D trick .. next to your face from a photo, focus on the overall black for a short time and you will notice that the background colors will fade
Based on this service, Italian researcher Giovanni Caputo, in a 2010 study of the experience, involved 50 volunteers who asked them to look up at a mirror placed inside a dimly lit room for 10 minutes. The results were amazing:

- Of the total number of participants in the experiment: 66 percent said that they observed changes and anomalies on their faces.

- Out of the total: 18 percent said they saw animals such as cats and pigs.

- Out of total: 28 percent saw the face of a stranger.

- Out of the total: 48 percent said they saw monsters and fictional creatures.

Of course, this trade despite its fantastic results, but it does not fully explain the myth of the mirrors, for example, 66 percent of the participants said that they noticed changes and anomalies in their faces after a period of concentration of the system, this can be interpreted according to the disappearing environment trick above, since the concentration of the participants for their presence causes a gradual lack of interest in the brain and is used for the rest of the image that led to the edge of the face away and live, but that we can explain the vision of some participants, animals or people vaguely with the appearance of spiteful and evil or objects , ugly and scary. In fact, the researcher Giovanni himself does not find an explanation for these visions, but he believes that this is associated with consequences and psychiatric ones. These few of us are facing a real dilemma because this study, instead of scientifically and logically explaining real myths about mirrors, it grew out of the loss of us and made us wonder: Is this really just an optical illusion of decimation of the virus? ? Or what is actually after the other, hidden inside the mirror, may come to wonder to find us in the abyss of the terrifying.

Finally, dear reader, you can also try your luck with the mirror, reduce the extra space, and then look intently at the reflection of your face in the mirror for a few minutes .. I hope to hear from you the score and what you see is, of course , if you don't show you Bloody Mary, tell you and you with them from the inside of the mirror will lock you there forever! ..

Teleportation .. Between fiction and scientific truth

As long as people dreamed of access to science and ancient wisdom, and they devoted their lives to finding and choosing all the ways to realize the hidden truths that induce the catacombs of oblivion, darkness, radioactivity, some of them succeeded civilization in the speed of detection was the reason for their prosperity. peoples. However, some laugh at these things and consider them just fairy tales and legends unfounded, but that over time, scientists and leaders, they do not know little, and there are so many things in this world still hidden from them, and said, that Isaac Newton once said: "You, as a child, playing on the seashore, I amuse myself by finding a pebble or a shell better than usual, while the perimeter of truth is not revealed Big before my eyes."

Today we learn about the phenomenon of instantaneous transition (teleportation), which is one of the phenomena that sages talk about and how does it happen? they used paper that way, and then lost their secrets, and no longer identify only coincidences, people do not know about anything and did not realize it before, but, as they say, shock is only a path from the paths of truth.

The Tale of Jil Perez the Alien

Instant transition between imagination and scientific method.
Suddenly appeared in the center of the Government Council in Mexico
The resulting story of this guy, largely from historical sources, remained controversial for hundreds of years.

Jill Pierce, the guardian of the emperor and the Spaniards, who lived in the 17th century in the Philippines (Spanish colony), in particular in the capital of the country, Manila, and who followed the regiment of guards entrusted with the protection of the governor's palace in Manila, was known for her dedication and strict discipline in her work. ...

The morning of October 24, 1593 was not a normal day in the Philippines, on the seventh day there was a murder of the situation by pirates, and it was not a normal day for Jill, too, it was one of the strangest days in his life, when he stood, as usual, in shift guard and felt a bit of a drowsiness and fatigue from the hard work of anxiety case the country had most experienced, so I just dropped my eyes to the brightness of the dream and closed my eyelids for a few seconds before opening my eyes again .. he found that stands in a different place, unfamiliar to me, which I have never seen or visited in my life, completely different from the point of view of building models from Manila .. He felt around him his guards, dressed in different designs and colors from those that wore ISIS, of course, these guards are also like them, from the closet and the generation that they came to us and asked him about his identity, he told them that he was guarding the situation in Manila, they followed what he said and thought, th some of them are drunk, this is not the governor's palace, it turns out that this person is a stranger, the body of some Filipino is there ?! Because now he was standing at the very center of the government council in Mexico City, the capital of Mexico ..

It was a hard shock for a generation, I swear to them that he was standing a few seconds ago in front of the governor's mansion in the Philippines, and he took a moment's nap just because of the stress at work to a clear maximum in the capital city of Manila, due to the rape case, but no one I did not believe him, and indeed with them, because as seconds go from one country to another, separated by thousands of miles. And with the two countries belonging to the emperor and the Spaniards, it was difficult to verify the claims of a generation at the time due to the difficulty of communication, there is no phone or email, and the news takes months to get from Manila to Mexico .. that's why you planted the next generation to jail to be brought to trial.

And when he was brought to court, he told them his story, told them the information that the governor of the Philippines was killed, but no one in Mexico heard this news, she was re-imprisoned.

Two months later, an irrigation business moored in Mexico, came from the Philippines, and, through them, the news of an attempt on the situation in Manila arrived, trust the new generation with a gesture, as one of the sailors on the ship, a Filipino swore that he knew Jill, saw him repeatedly before Governor's Palace in Manila, let go to the surprise of everyone who heard his story, and returned with the ship to the Philippines.

Psychic Carlos Mirabelli

Instant transition between imagination and scientific method.
Mirabelli was a psychic .. and shows in the picture on the grounds that he flies through the air
Carlos was the son of the city of São Paulo in Brazil, was a psychic, claimed to have lived a previous life (reincarnation) several times, and it was said that he could speak three languages, including dead languages, and knowledge of science and spirituality, ancient. in fact there are too many novels about the abilities of Carlos Hax, what does this mean for them there are those that talk about the development of Alanya, where the witness is not someone who saw him, that he can compete in seconds from one place to another, it evaporates like fog in the air to show elsewhere.

This is a story about many not that his friends witness the realization of the train station minutes after you call them to tell them that he arrived in the city of Sao Vincent, that it was far from Sao Paulo to 56 miles.

The accident of Rudolph Finn Orbitz puzzles

Instant transition between imagination and scientific method.
Suddenly appeared out of nowhere, in the middle of Times Square
It is said that one June day in 1951 in Times Square in New York, USA, a man in old-fashioned clothes and a strange physique suddenly appeared from nowhere, appeared so suddenly in the middle of the street, and he was so shocked in all his amazement. he looked in horror at the cars that dripped around him, and when he tried to leave, heading towards the pier, he was hit by the car's accelerator and died.

The company obtained the man's body and searched her mind to find Wawer's belongings in the personal name of Rudolf Finn Orbitz, which were reported in 1876. The police did not find anyone with that name currently living in New York, but the investigation came to a guy named Rudolph Finn Orbitz Little, and he died five years ago, moreover, asking his wife, told them that her father of her husband's name Rudolf Finn Orbitz suddenly disappeared on the street in 1876 and never reappeared, and in the photograph of her husband's father it turns out that he is the same person who received the company body, it disappeared in 1876, only to reappear in 1951.

Of course, this story thinks about the folk tales that are flowing through people, and there is nothing officially confirmed, and if your health is indeed, this, in my opinion, has to do with time travel more than you know, moving in real time with it. the included disappearance of a person and his appearance again in the last place with a large time lag.

Philadelphia USA experience

Instant transition between imagination and scientific method.
Experience the mystery of an alleged United States criminal
They are reportedly undergoing confidential litigation by the US Navy to test live aspects of Einstein's famous Unified field theory.

On October 28, 1943, at the port of the Navy in Philadelphia, trade was conducted on the USS (USS Eldridge) and its crew, thanks to the release of an electromagnetic field of very strong destructive effect.

According to the novels, the ship disappeared amidst the green mist, only to reappear 300 miles later in the port of West Virginia, and then not select again for the show in Philadelphia.

The experiment was partly successful, but the impact on the sailors was devastating, with some reported reappearing and exchanging their bodies with the ship's metal, and the entire crew later suffering from health and mental health problems, prompting the US Navy and government to speculate on trade carried out to hold them.

Youtube move real

Did the car appear out of nowhere, I saw it myself.
There are many videos of the transmission incidents of the alleged capture of Annie by the camera, and there are also several interpretations of these passages between certified websites .. we will not judge here, but we will go to one of the most famous sections labeled drive The Ghost, about a car suddenly appearing on the street at one of the crossroads in a Russian city.

Instant jump scientifically

Instant transition between imagination and scientific method.
Are there material cleaners that have been from one place to another?
In fact, at present there is no way to instantly transfer an object from one place to another, and there is no instant transition only in the plot ideas of science fiction, but scientists until today have not been able to transfer information only at the level of the amount of information infrastructure (qubit) in the known as a quantum of teleportation, that is, they managed to transfer the amount of information infrastructure of corn and the corn itself.

There are experiments conducted by some research centers around the world about the possibility of an instantaneous transition, but there is nothing concrete yet, only ambitions, theories and statements have not been established.

Debt and relocation immediately

Instant transition between imagination and scientific method.
Belkis the Queen of Sheba .. and her famous throne All those familiar with the Quran teach that there is a surah called "Ant", in which there is a sacred verse, in which Allah says: (he said that with a note from the book I went with him to perform party to you). Of course, this verse is about the throne of the Queen of Sheba, who asked the Prophet Solomon to make it right.
Of course, different inspectors about the person listed in the verse as "The Book of Knowledge", there are stories that this is a goblin from goblin sex who mocked his God Solomon, in another person's novels to be a good person, his name is " I'm sorry for the clear signs in Mecca. " In general, this is not a search for a store, we are here, what is the meaning of this topic is this supernatural ability to miraculously bring a bride from distant Yemen to Palestine at the moment does not exceed the blink of an eye, no, I'm talking about a moment, but this is the same speed of timely implementation of any maximum at the speed of light.

Of course, there are narrator stories for good people or Halloween in all religions and sects that have had their merit or miracles associated with the transition point, in the Sufis, as there is a property known as the "folds of the Earth", for they are said to have been able to cross in seconds, from place to place and from country to country in the blink of an eye.

In Spain, the monk Maria de Area became famous in the 17th century for his ability to teleport, where he allegedly contributed to the rapprochement of Christianity between the Indian tribes of Mexico and the United States, without moving from Spain initially, and he was covered in the blink of an eye by the monastery in which he lived from Spain to America to his glasses with gold crosses and wrote about Christianity.

What is the history of the mysterious lights in Hesdalen?

There are several areas in the world where strange phenomena have been observed for decades, without scientists being able to accurately explain these phenomena mentioned on the site, the most famous of these phenomena are to see the lights of mysterious bodies moving in the sky, the phenomenon of crop circles, to find animals and cows killed and distorted by a harsh and very precise control, as if she took samples from her body using a complex mysterious technology and then threw it away .. flying saucers .. Which has not proven its existence, but inevitably to many around the world, our today's article will is dedicated to one of the most famous and mysterious strange phenomena in Europe ..

Between 1981 and 1984, there was an increase in the frequency of sightings of strange anonymous lights over the crook valley in Norway. More than 15 attendees per week were reported. Although the phenomenon of the mysterious aurora is not new in the central region of Norway, where it has been seen since the 1930s, the species were not nearly as dense. Since 1983, he has been involved in ongoing research into this phenomenon by the scientific community, academia and government agencies. Despite continuous research for many years, the emergence of many hypotheses is trying to find an explanation, however, there is still no convincing explanation for this phenomenon.

Balls of light are bright, unknown flying objects and animals

What fairy lights are mysterious in his rogues?
Photo taken with the lights of a mysterious region
Its employees refer to the area adjacent to the Valley Championship, 12 km and 120 km south of the city of Trondheim. The population of the country and neighboring countries is 150 people. And the region as a whole is a mountain located at an altitude of 617 above sea level.

I began to notice the phenomenon of lights on a late 1981 scale, when he started the people of the valley, suddenly seeing many mysterious alien lights in the sky. Mysterious lights appeared in different parts of the valley.

Hundreds of highly bright photovoltaic balls are observed at different times of the day and night. There were about 20 participants per week. Residents report the involvement of photovoltaic objects in the Up and Down Movement, sometimes very close to rooftops. These lights appear in different shapes and sizes, brightness and saturation. In some cases, they changed their forms during the move. It has also been observed that some of these lights take on the shape of a large ball about the size of a house and are very bright, and may retain their intensity for the former.

The speed of these illuminated objects was varied between very fast and slow. Some of these objects are reported to be very slow approaching the surface of the earth. The faster the movement was measured by the team from the University, the faster it reached 30,000 km per hour. There have been attempts to test and measure other scientific objects. Some have pointed out that the intensity of light reaches the MW unit. Scientists have found that lights react and react to other light sources, such as a laser beam, by blinking and sometimes amplifying the ground.

According to the residents of the region, participation and observation of other types of objects was also carried out, where they said that they had witnessed the heaviest aircraft of a large metal shape with various cars, cigars and a cylinder .. moving at different speeds, these performances are in many ways similar to performances classic UFO convention about UFO sightings in different parts of the world.

Some of the locals also reported seeing large, rectangular, unexplained holes in the ground, created by cutting off a large portion of the wet grass, which adds a ton of weight and puts it several meters away. It was noticed that the roots were trimmed with a special type of code. And the chopping off of the roots of trees in the grass and the formation of crater-like, largely crop circle phenomena, which are also associated with flying saucers.

In fact, there are reports of stranger and more mysterious things that happened near his workers, especially in the case of animal distortion, which is similar to the phenomena of animal distortion that occurred in other parts of the world, where cows and other animals were found killed and mutilated in an unusual way. resistant to attack by local predators. In some cases, the lights of mysterious flying objects or distortions of animals unknown in the scene are visible.

And still no explanation

The video clip of the documentary program details this phenomenon
Many explanations and hypotheses have been proposed in front of researchers and scientists, but none of these hypotheses provided a convincing explanation or clear to the views of their crooks. Some of these hypotheses are not fully understood, for example, the connection of this phenomenon with a ready-to-go self-awareness is associated with the presence of and - a rare metal that exists in Scandinavia - self-awareness. Another hypothesis suggests that this phenomenon is caused by a more complex process in the atmosphere, such as the ionization of air and dust.

Further research

This phenomenon is indeed receiving a lot of attention from many actors, including the authorities and the media, the scientific community and tourism. Several scientific colleges and universities in Norway are conducting their own research in an attempt to explain this phenomenon.

A research team from Østfold University monument surveillance camera to observe reality 24 hours a day, seven days a week, and has already collected many images and videos related to this phenomenon.

Flying saucer bodies in Norway and Sweden have launched a project called Project is crooks to study this phenomenon. There is another research program called EMBLA, led by Ostfold University and Italian National Studies, started in business since 1998. This program is aimed at studying the behavior of the electromagnetic phenomenon of light using the developed measuring instruments of the radio frequency spectrum.

The crooks of that time no longer have known places for those who are interested in part of the flying saucers, whether the plane was mysterious. There was work on a documentary called "The Gate: The Phenomenon of Lights in Its Employees." "Many books have also been written about this phenomenon.

Until now, the phenomenon of lights flying in the sky over the Gorbunov valley remains, but the species today are much smaller than what happened 20 years ago. But despite this decline in views, Crooks keep it is an important place to study unknown weather phenomena (UAP) and odd flying objects in the world.

Pyrokinesis: harnessing fire with the power of the mind!

No matter how long I have inflamed our imagination with myths about superpowered beings with extraordinary limitless abilities, perhaps these are not just legends, but, perhaps, hidden in the folds of her powerful abilities, are hidden in one aspect of the human psyche.

Pyrokinesis: the magical ability of the mind!
The mentalist Daniel Douglas Home ... one of the first to claim to have the ability to generate fire Of the capabilities voiced by the famous novelist Stephen King in his novel "the first" with "pyrokinesis", any ability to shoot and control it over a decade, this the ability is considered part of the telekinesis ability.
Or it can be defined as: "the ability to accelerate the natural vibrations that occur in the atoms of water, and the change in temperature, possibly to the point of ignition, if the material is ignited by the power of thought or will." And note that this ability has been studied by scientists for over 100-150 years.

In 1882, a publication was published about one of the local children. an article appeared in a local newspaper about a young 27-year-old student. This guy named William Edward became one of the famous people in the area through a generation of people for his breath! ... He considered his possibilities through manipulation with his hands, then he takes a handkerchief, which someone may be close to the mouth, and then rubs his hands vigorously as you develop, and immediately fire to your homes. "

Of course, there are those who now believe that Underwood did not feel the fire caused by his superhuman abilities, but he hid a piece of phosphorus in his mouth, and then spat it out for houses, used the warmth of his breath and the heat from rubbing his hands against each other. to ignite the phosphorus, which causes a fire.

Except that this claim is weak, since phosphorus is a highly toxic substance, and the moisture in the mouth of the process kit bought it if it was the kind of least toxicity.

But now let's talk about the controversial Filipino girl in the local media, "Emma", 3 years old from Iloilo, Philippines. He is able to shoot as soon as you utter the word "fire!"

Where, if I mention something that you want to buy to start a fire, appear out of nowhere and are consumed immediately. I mentioned the mother of the child in one of the media that: "everything that is consumed by my daughter, thinking that she will burn, she will immediately burn. You should not believe in such things, but I myself saw how it happened."

Pyrokinesis: the fire ability of the mind!
A child is able to kindle a fire as soon as you utter fiery words!
The father of the child, who declared "Benedicto is my pill," is stunned: "I cannot explain what is happening to my daughter."

She said: "We are drawing a housing network of disadvantage", which is a witness of the child's legal capacity, who also suffered this legal capacity, she said: "On February 21, 2011, I saw my daughter ejaculate near the kitchen, where they slowly moved from the gutted kitchen ceiling in the following days. people behind the walls! "

When Emma had to say the word "fire", meaning the radio, or the pillow, or the blanket ... it worked. Even our clothes in the closet were unsafe. "

On the cover of this strange story, reporter Caroline Village and Mark Villa Toulouse said that Carol said, "I saw the people who made Amy, even her underwear is on fire too. I don't think it was just a trick. I know that they are just poor, they cannot afford to burn those important things that they cannot get easily. The camera I was using suddenly turned on after I talked to Emma about people. When I checked the video there was no fire There is a tape, and then I tried the registry, but still there is an error on the display. But when we got to town, I tried to register, then everything works without any errors! "

While Marc Villa Toulouse further details: "The police have followed the events and wants to follow her, despite the fact that Emma is not doing anything out of the ordinary."

I extended my expertise, Amy even went to one of the pastors who intervened, claiming that the devil, who meets the girl's body, was responsible for the fire.

Pyrokinesis: the fire ability of the mind!
Fire where Amy decided I tested Amy, as well as the house you live in, in a rigorous scientific environment, '' said Bonito, Chair of the Research and Research Department at the university, until he challenged my abilities Amy with some experts: "it is inexplicable, it looks like a supernatural phenomenon."
It was used in testing thermal chambers to measure temperature in the area of the accident, and body temperature. Amy reported that everything was normal and there was no difference or temperature change there.

While Hell Victoria is a supernatural expert, Amy has the ability to ignite people! Perhaps acquired in previous lives, as accepted, where the belief in reincarnation reigns, some classes in the Philippines have.

Hypnosis and its Secrets: Science of Scientists? Or some kind of magic?

Messmer prestige enters with his usual cursory glance, the entrance is categorically breathtaking, dressed in his black cloak, pregnant, dick metal, dims the lighting and says: "Raise your index fingers like this ..

I want you not to think about anything other than your fingers ..

...

I turned the fingers of the strongest hand, which she pulled towards her automatically ..

...

I did stick my index fingers, and no matter how hard you try to separate them, it won't work .. "

Glue a real miracle to your fingers, just as Messmer pointed out to them, and he turns everyone in a panic.

John Messmer brought them to fruition with his fingers. And this is what happens. When you selected Messmer as one of those present and asked him to look at the metal bar that raises it high, he repeated a few words and then whispered in his ear, abruptly: "sleep" .. tea plunged him into a deep sleep.

Messmer (opposite the screen): "you will wake up from time to time, and that if I scratch my head, I will be your coat .."

Wake up screen Fischer Messmer talks and what scratches his head until he meets beautiful people and think their faces look out of curiosity. Vince took his coat and walked towards Messmer, squeezing his hand and the drink in it, in it.

Messmer: "I don't want to"

The young man says and finds himself on the verge of tears: "the weather has become cool to put on please," Messmer replies and shares his knowledge, signs of satisfaction and pleasure appear on the young man's face.

Messmer (asks the young man to the applause of the audience): "Did you put me in your coat?"

The young man appeared in complete amazement and embarrassment, he and May got pink cheeks that made him ashamed: "I don't know ... it was just a stupid idea that suddenly occurred to me and I completely took control of myself."

Make the audience laugh and forget about the curtain

You must be wondering, who is it that determines the development of change? And what is practiced on people who are so disobeyed that they pass to such a degree as to make themselves a laughingstock?

There must be some expectation that the challenger will or maybe stop, some of them are sex and demons. I didn't miss and I'm shooting at them.

Secrets of hypnosis: science and scientists? Or a hit of magic?
Franz Messmer .. leading hypnosis in the modern era
His name is Franz Antoine Messmer, he is not so intimidating about the mission of peace in the soul. They are bidders, a physicist, and a German doctor. He excelled at representing hypnosis as an acrobatics player at the age of fifteen, and inherited this art from his grandfather. He begins his presentation on the routine of test fingers to be trained from among the masses of people with the greatest potential for diversification, and then goes all the way to what I explained to you at the beginning of the article. Even now, his life was calm, stable and happy, even he majored in medicine and spent most of his life studying hypnosis and its secrets, as safe as its therapeutic and diagnostic properties. It is Messmer who depends on a piece of Magnet in the treatment of patients, where he conveys the theme of pain and is also used by way of originating in the consciousness of patients. A safe system that has a liquid magnetized in the human body, a similar liquid contained in certain minerals, when used correctly, can affect the human body and brain. From here, he launched Messmer into this process as "hypnosis" and some said that the origin of this recording is an instant dream through it of a person who placed it as a magnet attraction

Its origin and history

Secrets of hypnosis: science and scientists? Or a hit of magic?
Rose hypnosis in the minds of people in prison and reviews
Hypnosis is one of the most ancient sciences, not an innovation of the Messmer, but as long as it existed before his birth for about three thousand years. The inscriptions on the walls of the temples of the pharaohs, which have survived to this day and clearly reflect hypnosis sessions, are the greatest evidence that the ancient Egyptian civilization was a pioneer in its use. It was received by the priests to strengthen the faith of the devotees and even in the future. They also said that his homeland is India, where the poor used it and recorded it in their books, and then the Greeks and Babylonians used it. Having spread in some countries and extinct in most countries, it flew into oblivion for many years due to the status of its business to many of the sciences. And why not consider them only through Messmer in the 18th century, where they were tired of the European princes, and some called them the "marginal" percentage. But soon they did not deny the eternity of our hero, and the last years of fame and fame, where some believed that what he was doing was affected by magic and witchcraft, to the point that the medical organization in Vienna did not consider him a study and even refused him membership , which made him leave his hometown in 1870 for France.

There appeared researchers of piercing, in their theory they, Lavoisier and Franklin Benjamin, confirmed that the liquid that Messmer spoke about does not exist and that this process is only the power of suggestion, and since then it has changed the concept and received the name "hypnotherapist"

Secrets of hypnosis: science and scientists? Or a hit of magic?

American doctor Milton Ericsson and invented a new way of hypnotizing took her period away from the old road view

Since 1923, the American physician Milton Ericsson appeared and invented a new way of hypnotization, which removed it from that period and more out of respect for the human self from the old way, where you do not say to take direct orders and not on the vertical relationship between the hypnotic and the processor. Much of Ericsson's research is reliable to doctors to this day.

What is hypnosis?

Secrets of hypnosis: science and scientists? Or a hit of magic?

After all, a magnet is a special case for the sense of time and external factors lost in it. Various scholars and experts on the precise definition of diversification. Developed several theories and refutes another. And not a single theory is finally confirmed. But there is one definition that everyone agrees with, and whether it is a case of desertion of consciousness or a state of consciousness, does a person under the influence of hypnosis and his feelings and his memory mean only at a time when he is part of the critic and accountant of consciousness. Features of the human mind through this experience of unique sensitivity to high command execution with loss of sense of place and time and changing standards

The human brain in a normal state (wakefulness) is able to focus on several things loaded by it and cash them all, and the reaction at the same time in the case of hypnosis will not focus on only one thing, namely on the Voice of the hypnotist and sentences that will be unable to analyze or criticize them and therefore will be more accommodating and responsive to the commands of the hypnotist.

Confirmed by experts that we all go through this unique experience without feeling like when you drive your car for long hours and you deserve for your favorite songs, you will not feel bored, you will lose your sense of the slow passage of time. you will not feel tired of the length of the distance and traffic damage because your focus was on the music that took you by default everything, I was hypnotized by you without feeling

There is another example, more similar to the one when you sit in front of the TV and watch your favorite show, which you have been waiting for since when you furtively thought to avoid the problems that bother you - and moreover, the problems of life - and in this moment of wandering, and someone comes and asks you to turn off the TV, change it automatically. And what should you do until you find out about yourself and ask yourself the question: "What happened to me? I didn't turn off the TV?"

Secrets of hypnosis: science and scientists? Or a hit of magic?

People can hypnotize themselves in the same way .. there are sites offering guidance for this, there are YouTube videos claiming to be able to hypnotize you by focusing on effects like the above picture and hearing the voice of the hypnotist and he says you are look And there are a lot of similar cases and the fact that we go through and show that everyone is able to hypnotize themselves and surrender to the phenomenon of diversification or automatic self-hypnosis in return, not everyone can hypnotize another person without the fact that there are some leadership qualities, like a car and being with someone. that of steel in addition to being hypnotized by science.

You should expect some kind of tape in the person who will be hypnotized you should be a believer in this process and want to fight it in fact the applicability of hypnosis varies from person to person and according to the World Health Organization, 90 percent of the people in the world are diversified and usually children are usually over BS

Applications

The light of the connotations that an individual will receive, and under the influence of hypnosis, can change his behavior and inclinations and help him nod in getting rid of pain, physical and some habits that are locked in the phase and its minuses, but he cannot keep it to himself.

A video clip of the process of hypnosis is real in one of the programs dialogue arab
It also improves the emergence of students' attention levels and reduces anxiety and fears about succumbing to testing important material or gratitude for their fate, how to get rid of some embarrassing habits like nails, stuttering and panic attacks of all kinds, and help smokers to get rid of. from his desire to smoke, or one of the cannabinoids through entering the patient's inner work and fixing the risk of addiction in his mind and in such a state of serenity and relaxation can also be hypnotizing to get rid of some skin problems associated with the psyche of the individual, from routine and pressure

Confirming psychologists claim that inside every human being there are tones of different sounds and when a person becomes infected with an imbalance of personality, his vocal cords are affected, he loses control and can make sounds that are strange, scary, embarrassing and funny, like a woman in a rough voice is closer than the sound of a man to to a woman or vice versa, Yusuf Jalal explains some of the symptoms of demonic possession during sleep: "I got a woman from Alexandria who suffered from a strange voice coming out of hypnosis sessions until she was completely healed."

Robbery video with hypnosis
It has been proven that hypnosis is also effective in treating diseases that are part of intractable, such as diabetes and blood pressure, and also improves sexual performance and obesity, where he believes that the treatment of obesity lies in the brain, which controls hormones, hunger and satiety, and can even treat hot flashes that some women experience as a result of interrupting their lives.

Dear reader, please note that the patient's treatment is not an accumulation of drugs that are not devoid of serious side effects in the long term, since much depends on the person's psyche and the strength of his will, the more his mood was lifted whenever the disease was overcome faster

Hypnotherapist in the operating room lobby

The use of hypnosis since the nineteenth century as an alternative to anesthesia by injection is traditional, and that because of its analgesic effect in addition to its positive results and purity in the same patient, it is not done only with the consent of the patient and the surgeon, and also after passing the acceptance team, it is necessary to be comprehensive in hypnosis and after the testing procedure in order for the patient to remain under the influence of hypnosis for a long time

Secrets of hypnosis: science and scientists? Or a hit of magic?
The use of hypnosis as a means of identifying patients began with the emergence of this field in the mid-19th century.
Created in a special factory in Tunisia in 2016, the first brain process under the influence of hypnosis is the most serious process and took almost 6 hours to be successful, you will be the first Arab country in Tunisia and the second country in the world after France to rely on pain relievers the effect of diversification and now rightly in Tunisia they do 6 operations on the brain and two on bones under the influence of hypnosis and without augmentation, and now this method is used in 30 percent of operations in Europe and it is worth noting that he cannot hire hypnosis just to during certain operations, and not all of them recall the processes of the thyroid gland and the setting of the chest calendar, as well as varicose veins and treatment of burns and Gabr Alex and dental surgery, and obstetrics, and cosmetic surgery (suction to correct a crooked nose ..)

It is not entirely clear to understand the connection of hypnosis to feel pain without looking ahead to the dilemma of the biggest, and this is how the human brain works when exposed to pain

There is an area in the brain associated with activity significantly during the shedding of an external painful factor on the body or during the imaginary exposure of a traumatic experience and is the organization responsible for feeling pain called the cerebral cortex in the front of the brain. During the nucleation and production of large amounts of [Meredith] endorphins by the brain, natural pain occurs and is an analgesic that affects the activity of the frontal cortex of the brain, reducing its activity and, therefore, canceling the innate feeling of pain.

Hypnosis review

Loop hypnosis is your Goto talent, it was hypnotizing from the start as part of the program review
Hired hypnosis also in theaters and on the streets and in the circus in the form of entertainment, and also contributed to the birth of the review in the definition of this flag, to return as a unique and painful diversification of the Medical in the eyes of people, where they do not take it seriously and do not believe in the effectiveness of his treatment

In the field of criminal investigation

Think that hypnosis is the way of modern scientific criminal investigation and state interrogation of witnesses and suspects to obtain facts, can raise people to say in case of awareness or make them remember some details that could not be remembered in a state of awareness and recall some statistics, which among 50 cases allowed nucleation to get 60 percent additional information and think that many developed countries this means in response, including Sweden, where integrated hypnosis training in its educational programs and approved by the court in the USA in 1926 is fun to use to prove your innocence

In a California chuan in 1976, an unidentified man attacked a bus at gunpoint, reported the abduction of the driver and 26 children, and then put them in a transport truck. Fortunately, the abducted man was saved from escaping, and after interrogating the driver, it turned out that the latter did not remember the qualities of the driver. After you learned hypnosis, you were able to recall some of the bride's numbers and her qualities that helped the police arrest the criminal

Secrets of hypnosis: science and scientists? Or a hit of magic?
You Can Use Hypnosis To Get Information In Criminal Cases Enraged doctors and mental health professionals could be right off the bat about the American hypnosis company and the police using the drug, and giving some victims of violent crimes like rape the latest case details can cause harm to the best and aggravate their suffering with additional suffering and injury caused by the incident. As some militiamen fear, suggesting elements of the answer, this is either unintentional during the operation, especially after the discovery in the summer of 1980, in the city of Elena, of the body of a 14-year-old girl who was abused and killed as a result of a choke. And after questioning the three teenagers who had been with her last time, they said they were playing a sex game and abandoning victim practice with one of the boys their age, named Billy. And when he interrogated the latter, he looked tense and worried and emphasized that he had forgotten the details of that last evening with the dead girl, which strengthened his doubts about this.
The company decided in the end that the subject of children's diversification to improve the mistake they might have from some information after the delivery of the will and the detective confirmed to the Billy family that hypnosis would help reduce anxiety and tension in order to convince them of the idea, at the same time, draw the child's attention to the police are waiting for convincing evidence to charge him with a crime

During the hypnosis session, covered by the processor, Billy's body was covered with a lid and there they assured him that the operation would help him, and then they asked him to stretch out, close his eyes, take a deep breath and relax, while maintaining his concentration then the Wizard invited Billy to look back at the last an evening with a blackboard immediately admitted that he had to silence the breath of Posada's victim of death

There was no evidence to condemn Billy other than his confession under hypnosis and the incentive that forced the rest of the boys and the victim's refusal to practice sexual play with him, the coroner stated in his medical report that the victim could not have been her death caused by Posada's suffocation and the chest scratches were not can be called by Billy

And from there, the defense argued that the child feels Billy, that he is a murder suspect remained firmly rooted in the subconscious, and this is what caused the child to fake his confession under hypnosis and perhaps offered him a way that was supposed to be behind the crime and is weak-willed, his imagination said, in order to invent all these events without resistance in addition to the pressure the police put on him, adding to the child at this age several changes in the physiological and psychological development of Fisher, and the girl abandoned Billy, who was stabbed to death in his masculinity that the universe has some grudges we showed this in a hypnosis session maybe he wanted inside him that he was killing her, but he did not dare to do it on earth and in this case Invented some memories to tell about his masculinity and some of the dangers of hypnosis is to lie to a person and then believe his lie as if it were real.

With regard to the concepts of criminal proceedings, the issuance of an opinion or certificate of free will and conscientiousness of the whole is required, as well as agreed international and national legislation on the right of the accused to refrain from answering questions from investigators - this is what is called the right, but this is the right of the accused to lie in for self-defense.,

The reasons why some developed countries categorically prohibit the use of this method as an encroachment on the freedom of the accused and a form of torture or service, including France and Germany

To risk

The existence of some diversification of risks depends on specialists, as well as on the whole of science. Can cause sometimes narrow fetal formation, dizziness, drowsiness, headache, making fake memories completely unreal and even some BDSM

The look of the hypnotized ended in tragedy

Secrets of hypnosis: science and scientists? Or a hit of magic?
Astrology: A Spiritual Footprint by Troy Cable and Michelle Dusty's Stuck Astrology in their book Astrology: A Spiritual Law tells the story of a woman who knew her life destroyed by hypnosis, she says: "she was a child, a beautiful blonde, her face is like the moon it was all her fault that I went out in the company of her beloved father to watch a hypnosis show, where my father volunteered to serve the craft in front of those present and freely gracefully covered the man in deep sleep so that he would make a decision that echoed in his ears: "you do not know this girl. this girl is not your daughter, you can come alone and not know anyone from those present .. Go and ask her about the parents "and then finished the show after tearing the girl in tears among the laughter of the audience decided sleeping pills to finish the show at last there is a person to wake up and repeat the process several times, but my father's person woke up this time and settled in the seat and started cursing and playing at the moment when you think about doing such a job, and then yelling at the audience: "they called an ambulance help."
The implacable man was taken to hospital for 24 hours before he regained consciousness, and then returned to his family. And after a few months, he completely changed his behavior, surprisingly similar to hypnosis, perhaps having received a fierce beast from

him, until he became a stranger to his family. Become rape your daughter for nine years without shame and do not be ashamed of the Almighty and be in obedience to the will of your animal, and even his other children will not spare from rape. And it was said that hypnosis is a gateway in the other direction, forcing her to become infected with a demonic ... years passed and it became so that a child, a woman who was very funny worried, suffered from severe psychological problems, I proceeded again to commit suicide, like her brothers, one of them serving a sentence in prison, the question of confession is on his daughter. "

In fact, this is not the only force that ended the bride by hypnotizing the end of the unexpected all over the world, which makes some countries pass strict laws in this regard, where Belgium's warning, the practice of this kind of proposals is final under the law of 1892

It is not recommended for pregnant women and children to undergo diversification and not even those people who use stereoscopic medical and organizational means to damage the heart, where it can interfere with the magnetic field with the field of work

Unlike stationary dependencies, the same can be achieved for the mission, to solve them with all its problems, like a magic wand, in order to change for the better it is necessary to go through the stages of the basic first search and war, such as when the mountain climber rises, he had to work and practice and devoted his time to be able to achieve it, probably failed some of his attempts until finally arriving at the top, and he must feel joy and pride

The elite power that hypnotizes you can be achieved without going through all those stages that will help in building a balanced personality, strong to resist the blows of fate.

What is the real controversial scene in Can You See Me Now?

Secrets of hypnosis: science and scientists? Or a hit of magic?
The film combines mystery and imagination
Watch the movie The story of four people who called themselves "knights of four" provided entertainment depending on their juggling skills and some psychological tricks

One of the scenes shows the actor Woody Harrelson, who embodies the role of the boy Merritt Mickey in the film and commits an armed robbery for a couple at the airport, not having any weapons with him and not using him, catching his weapon with his ability to hypnotize, then enter consciousness. and read what happens in a second and what may be touched upon in another article. At the beginning of the scene, the world stopped the man and, walking around the circle attracted to him by the conversation, noticed that he was hiding some secret, a secret that could destroy his marriage. When the hypnotized wife then gave her a heavy tongue and they stuck to the Earth, which made her incapable of speech, and freedom In the meantime, the man looked confused and all the time thought about treating his sinner, which connects him with his brother and his wife to take this from Merritt the secret is that $ 250 was all he had in his wallet.

A wife made to get her husband back from a trip, the "job" initially did not work. Take Merritt reading his wife's mind loudly in the ears of his wife, who looks completely traumatized, but she reacts helplessly when I tell Merritt to her husband that he can make her forget the whole thing as soon as the little fighter offers to give him everything he had. Of course, this traitor did not dare to risk his marriage, he agreed.

Secrets of hypnosis: science and scientists? Or a hit of magic?
A scene that hypnotizes where Merritt's wife is
It is worth mentioning that the method adopted by Merritt in hypnosis of his wife is a method considered by experts to be a snapshot, fully measured, but then the question comes to mind, can a woman really forget such a dangerous thing as her betrayal?

The answer is yes, with some reservations, since psychologists claim that there is a hidden side of guilt inside every person, they receive information, refer to it and criticize it, and also determine the corresponding reaction not so worthily by such means as those wives who can infect very suspicious, I will not trust the words of my husband after that, going to look at it all the time, you can hate your sister who left them .. without being able to find an excuse for your actions.

You will learn to coexist with cheating without having to remember the type of obsessive-compulsive disorder that hypnosis can cause. This scene got the audience thinking: is it possible to hire hypnosis for dastardly expenses and blackmail with money?

The answer is yes, says psychiatrist John Stratton (can learn hypnosis techniques as well as they know medical surgery and years of training when it's the same for a change), simple but ethical considerations are often and skills are only acquired over time)

But think a little, dear reader, that it would be easier for you if you were a thief? Carrying a pistol and threatening the victim directly? Or is the alleged victim hypnotized? Personally, I would choose the first solution that will save me time and effort, and acquaintance with the art of hypnosis and its methods will require many years of training and indoctrination, in turn, will become the second option for filmmakers and filmmakers to endure such a special and exciting and precisely the filmmaker Louis Liter did it.

Critics say that the human psyche is much more complex than the picture presented by the film, where the victims of the four knights are covered with putty in the hands of a baker, put them to sleep and read their ideas, and reveal their secrets with ease, while these methods always have a percentage errors and therefore the success of psychological tricks at all hubs and bears in the film is only directly from the exit the excitement ends, and if it is based on scientific facts, then only if
If I don't understand why science fiction is "now you see me" I would share a few films that require sites to nucleate, and these oligarchs of unprecedented splendor have burst into dependence

Levitation: Take to the air between truth and fiction!

Over the centuries, many extraordinary phenomena were reflected in the hands of prophets, saints, spiritual mediums and in various human societies, exciting questions about our nature as creatures of human mysterious nature, about the cosmos in which we live outside the boundaries of a physical product, we cannot but learn about one of those phenomena, of which levitation is defined as a special representative, or what is known as the phenomenon of air, or, more precisely: "the height of the human body and other things in the air by the power of an anonymous spirit challenge gravity."

House of the Mentalist Daniel Douglas
The Mentalist Daniel Douglas Home
From the months that the psychic's ability was examined, the famous Daniel Douglas Home received his first ascent experience, when he was present at one of the spirit meetings in August 1852, he suddenly rose into the air, wounded chills from head to toe and felt a sense of awe and delight. ... He rose into the air a little and then returned to his place, and then climbed to the second, and the second launched high into the air, until he reached the ceiling of the room, where he touched the ceiling with his hands and feet slowly, after a period the bus began to control this ability taste and demand, it was believed that the spirits that visit are responsible for lifting him into the air, and in 30 years of practicing reviews before the General Assembly, saw thousands of people with their own eyes the ability of his curiosities to rise into the air, the most famous of the reviews conducted by the house was the presence from Lord he is Lord Adair of the County of Master Lindsay, where he hovered in the air and left, flying out of the window of one of the London high-rise condominiums and hitting the architectural interview window.

And I watched how Old English Mr. "William Crook" carried the earth many times, and confirmed that there is no possibility of deception in this process, once the wife of Mr. Crook, who was sitting near the house and who was brought into the air, got up! ..

Psychic uses country Arduino .. show it here and she will reject the table in the air
Psychic uses country Arduino .. show it here and she will reject the table in the air
A psychic from the famous "Asian country Marcelino" often settled down on the ground, and, being able to raise or lower the temperature of a substance, I have long been convinced of the authenticity and reliability of his special ability with the help of laboratory experiments of rigorous procedures of scientists outstanding at the beginning of the 20th century, and after the last use a special session told the mathematician and astronomer from the French "Camille Ger" that food is not a height from the Earth stranger than the phenomenon of the Iron Sector Magnet!

Adorable Indian shows love costumes flying in the air
Adorable Indian shows like costumes flying in the air
In the middle of the nineteenth century, the next "Lewis Jack Khim" traveled the chairman of the court in the city of "Chandernagore" in colonial India and throughout the country to

obtain more detailed information about the miracles that the "fruit" experiences (to two Sufis, Indians) wrote about many reviews, which I saw with her during my tour, but what interests us here is that I saw this phenomenon with my own eyes in the center of the "free", since one of the faces is called "covers" with the maximum lot in front of him, one of which has a height from the floor by 12 inches, and a teacher with over 8 minutes!

Other cases

Yogi Milarepa, one of the learned Buddhists, had such an ability that he could walk and sit while levitating!

The mystic Saint Joseph was born on June 17, 1603 and became famous for the fact that several times he rose from the ground for long periods of more than an hour.

Monk Magdalena de la Cosa (1487-1560) from Cordoba - Spain, she enjoyed this ability.

The famous photo of Anfield's release of a girl named Janet Obsessed is flying in the air
In cases of demon possession, levitation can occur .. this is the famous photograph of the Enfield number of a girl named Janet obsessed with flying in the air
Margaret Rolle, a little girl from Boston in the sixties of the last century, suffering from demonic possession. it was alleged that Peña got out of bed in the presence of many witnesses.

Saint Teresa, who lived in Avila in 1515, claimed that she rose one and a half feet from the Earth in less than an hour.

Of course, there are many people who have considered this phenomenon, but I think that this is enough to prove an important point - that this phenomenon was familiar in the traditions of many peoples.

In India, as many of the mystics believe the Hindus, the Indians created levitation, or what they call "flying yoga", which consists of three types in accordance with their faith (tennis - swimming - flying) and the mastery of constant contemplation of other abilities of another.

In Judaism, it is believed that levitation is a phenomenon of magic performed by invoking spirits, for example, it is believed that God mocked the prophet Solomon in everything, including in sports, which was mainly used for a flying carpet, regardless of weight. which you need!

We also remember that many magic books of the Arabic language contain many spells and talismans that gain flight and floor height!

Photo of Colin Evans famous
Photo of the famous Colin Evans ..
Despite the prevalence of this phenomenon, which has been viewed by people since ancient times, there were many phishers who claimed that ascending into the air, like

Colin Evans, where he appeared in one of the photos, he rose high from his chair in a dark room in front of a bunch of his assistants. except that Evans' trick was discovered where he performed the one who was holding him in his hands so that he was the same as in the photo and all I had to do was jump on his chair in the air and then pressing the button at the right moment shows how it flies in the air.

The conclusion from the question addressed to you, readers, is: do you see the phenomenon of levitation as real? Or is it just a gimmick that intermediaries test before people?

Camera: two in one

It could be any of the chimeras from the outside to know that this situation can occur in almost one third of multiple pregnancies, the condition or phenomenon is actually strange, maybe you have not heard that many people were incredible to see the possibility their occurrence. To show you the strangeness of this disease, let me tell you a real case that happened to an American woman named Lydia Fairchild.

In 2002, I broke up with Lydia for the sake of her husband, at that time I had a couple with two children, and Lydia is pregnant with her second child, filed a lawsuit for receiving routine care at the expense of children, the United States asked the parents to provide a sample of their DNA. in order to compare him with the DNA of his children, so as to make sure that he is the father of my children, in fact, by comparing the DNA, it turns out that the husband is already the father of two, but the shock was as a result of a test for the whole world, from where the result came identical, i.e. That is not so technically not a mother of two children, with all the birth documents, statements of witnesses proving that she did them, but is DNA lying? .. of course not . Therefore, the person of the court charged Lydia with the fraud said by her children and their welfare.

Camera: two in one
Lydia with children (pictured with her fourth child)
For Lydia, it felt like a nightmare, how not to be her, or the children who gave birth to them, was it something incredible? Since I need a stranger by all criteria, I come as a precautionary measure to be a witness in the middle of the yard during the birth of Lydia her second child, and that is, to take a sample from the mother and child immediately after birth for DNA comparison, was already a fulfillment all requests and recommendations of the judicial letter to settle scores, get out of fantasy and nuclear - for a child never coincides with the DNA of the whole world! ... This result baffled everyone, and the court will choose a new DNA .. This time, he decided to take blood from the babies from an unusual place, so they took a sample from Lydia's womb so that the result would perfectly match the baby's DNA, but how can this happen how humans have two types of blood ?! .. Thinking is just a camera phenomenon (chimera), in truth, Lydia was not the mother of the children in full, she really does, but they are also the children of her twin sister, initiated by Lydia when they were fetuses in their mother's womb! .. Thus, Lydia is the children's aunt too !! ..

That moment, what is this nonsense .. how to be their mother and aunt at the same time? .- Is Wayne even their aunt? Do you live in Lydia's body ?!

Yes, believe it or not, but it's true, you too can have a body with body parts of a brother or sister that you don't know anything about ..

What is the matter and what is the reason?

The chamber of a person or animal carries two types of cells or DNA in its body, and this case, when the twins are still in the uterus, where there is one case of eating twin brothers by absorbing cells, without causing any problems or symptoms in the mother, she does not experience In response to this need, this phenomenon is known as Vanishing Twin Syndrome and was first discovered in 1945.

Camera: two in one

In the first image on the left, there are twins .. a small second and, in the rest of the images there is only one fetus or the other has disappeared and was completely absorbed It is worth noting that the incident does not happen randomly, the embryo of which is swallowed or absorbed by the tissue of the fetus, the other will often be sick or abnormal, while the other fetus is in excellent health, and can be partially explained by the theory of natural selection, i.e. in the latter Since the proliferation of ultrasound devices, I diagnose this condition more and more often, for example, it is diagnosed when there are two fetuses in the uterus during the examination in the previous week or during the previous pregnancy, but when it returns, only one embryo does not appear in the jewelry next, which indicates the onset fading Gemini.

Often does not cause these problems and the feeling that a person may not know about his existence, when you do not accidentally as a result of medical examinations related to the disease, but sometimes it can cause status problems, not only in women, but also medical problems, like this happened with one of the blood donors to another person, where the body of this person who received the blood did not react to the donated blood. 5% of this version was presented, rejected by the body of the person donating to us.

An example of this case is also what happened in the 1950 Olympics, when I was a player in a Danish country, and by the discovery we find that some of its cells, male and other female, did not advance at the time to be removed from the Olympic games in general and pull out the rewards received, and were not cleared until after her death you examine several cells of her body.

Camera: two in one
The camera is a mythical creature .. shown in the picture by Tyler Mull .. her belly has split into two different colors
The term camera (chimera) itself was an expression of the presence of a person or object carrying two different types of cells in its body, while the camera is by origin a creature of Greek origin, it was called an alien figure with the head of a lion and the body of a goat and the tail of a doo body. This contrast integrates the physical and can be seen in humans by the camera, although in most cases it is invisible, but sometimes it is reflected on the body too, and not only at the level of blood and data, as if they were just born with two eyes of different colors (color difference eyes have no other reasons either.), or even skin color is different, as in the case of a singer named Tyler, philosophy, and the children who saw for a short time asked about the reason that the stomach is divided into two different colors, and after tests and reviews for a long time found out that the situation was caused by the presence of two types of DNA, there is something that was reflected in the tissues of the body, and this situation often occurs when the twin syndrome fades away in asymmetric Gemini, where naturally, in order to whiten the female, an egg comes out to use one sperm out of millions sperm, but in some cases it leaves the ovary to obtain two sperm, to be homozygous twins, and in some cases Bl The moneymakers say that one of the cells of the embryo integrates into the cells of the fetus to the other. This situation can be medical curious, especially if it happened in an advanced stage of pregnancy.The cause of the onset of diseases that affect many of them, such as: joint diseases, weak muscles, this case represents a significant problem in the status of a person, since some of his cells

attack cells of his other twin are extinct, which a person knows to make serious diseases difficult to treat.

Finally, this case (Twin Flying), representing a share of 33% (one third of people) according to the study of English, which is a very large share, dear reader.

Maybe I or you are carrying this suitcase?

Skinwalker Ranch, visions swinging between the threads of mystery.

The farm "Sense" (skinwalk ranch) was named after an ancient curse, the Indians said that they were at this place, and is a property sitting on an area of 512 acres located in the southwest of Ballard, Utah, United States of America was called the traceroute fear of taking the area 51 civilians, but why all this?

We will trace the history of the Gorman family, that small family consisting of Tom's father and Ellen's mother with their sons, and their experiences in this place that they had not heard anything about him before, and which are written by the inhabitants of the surrounding regions.

The farm of "semantic" problems oscillates between the threads of mystery
Jim noticed a stranger's clauses in a land purchase contract
Bought Jim King's father in 1994, at a reasonable price, because he needed repairs and restoration, the last couple of old ones who lived here left him seven years ago, he agreed with them on everything from the purchase, but Jim noticed strange points in the contract, the clause on the absence of drilling in this region only after permission from them personally, but never thought about it seriously.

Strange things, I didn't give them time to even think about why you put this item, she saw the family on the first day of the creature, like a werewolf, wandering around the house, and ran to her father's house, brought a sniper rifle and fired several bullets at him, but she was needles, or less effective for this strange creature did not feel it at all and ran away, this was the first strange coincidence, and perhaps passed unnoticed by them.

A few days after I stumbled upon a couple of strange creatures again and she is in her car, but this time she was accompanied by another one who looked like a dog, not a wolf, and he swam around without any sign of attack or danger, then he disappeared.

The next day I called Ellen at the Forestry Conservation Department for the forties and made up about her that a pack of wolves got out of the car, hung in that place and had to do something, but her application was rejected because Volkov was in this there is no area, she replied that the last wolf was killed there in the twenties or thirties.

The farm of "semantic" problems oscillates between the threads of mystery
The image of the creature you see in you direct and influence
She thought Ellen had it available and that the reason for these misconceptions is that this place is new to you, especially, and they may have seen things disappear and then reappear in strange places, which is not it. which you can only see in the person involved, he says. - so I decided not to tell her husband and forget about it, but the evidence that refutes her theory was soon, when Tom reached the evening that day, I realized that the equipment, which weighs several kilograms, landed in a strange way and broke, and it didn't give them little boys ... This incident prompted Ellen to tell him about all the strange things, and then the family realized that perhaps she was wrong.

And Tom repaired and modified, and brought his cows from the king of his former self, with the help of his nephew, hoping to settle here. the story of what he saw behind him at dinner that day and threw his cloak in a dark place at night, so Tom and his nephew check the condition of the cows on their first day of ownership, but they for seeing the lights look like the front cars, it was not the first time he thinks that there was an intruder, so I decided to catch him leaving with the lights that were left back quickly and I thought that the Tom that was in the car was broken the fence, but he noticed after closing that he was completely healthy, and then hinted that the lights and the sky had momentarily disappeared. it was the strangest thing. So I saw Tom with my own eyes with bare hands.

He prevented his nephew from returning to that place and now another incident haunts poor Tom, whom he witnessed a few days later, and in broad daylight something marked with a strange round fly near his house renewed with a sound similar to the voice of ways to stroke each other, and then suddenly disappeared, and they buried him in these views, in order to resolve the issue of this place with one of the local residents, who answered everyone that this place was cursed with an ancient curse, the Indians cursed the "meaning" and that a creature that looked like a werewolf that he saw was this fucking legendary creature.

The farm of "semantic" problems oscillates between the threads of mystery
The image of lights, strange and mysterious
Days passed and the quiet resolution of the winter cold and far away, and his phenomena were strange for this family where I saw Tom this time more than a triangle flickering in the darkness of the night of darkness, he tried to understand what he sees because of the organization of the hunting rifle, he saw a similar shape figure VI is cylindrical, but his stomach is empty and his strange aura stopped moving strangely and then the lights disappeared with a dizzying speed, he did not think that it was a spaceship, he thought that it might be a portal to another world - the theory of multiple universes- he also noted that what is happening here negatively affects his family psychologically, so he decided to start looking for a buyer for the farm, but before leaving he received a parting gift .. A curious case on the other, but she is the wildest of the previous ones .. On the morning of a cold day I came out to check his cows and found that there was no cow in the center of the herd, so he began to look for her on the horse, and know her fate until he found I was cutting traces of a modern spot on my show as I started to run. suddenly something special scared and the fact that some of the small branches break ..

Follow Tom's effects to the borders of the lot, but she suddenly stopped, as if a cow had swallowed earth or taken to the sky, just days after Tom decided to go out with his son to spend time together and broke up. walking through the center of his vast farm and seeing the son of one mother of a bitch in a puddle of small water, she thought she was stuck, he decided to call his father, like lightning was a small cow shell, but he decided to pull her out of the cow from the center of the pond, only to discover that her entrails were removed in a disgusting and well-done manner at once and without spilling a single drop of blood.

Farm "meaning" the problem of fluctuations between strands of mystery
The image for one of them Tom chose her womb completely, and oddly enough did not stir up the blood on the cow
Tom said that he lost fourteen cows in his final days between disappearance and a mysterious death, before he could finally sell the property in 1996 and bought a new farm about twenty four kilometers away, but he did not finish his old place, he wanted to find answers with new owners who would sometimes come to make his cows. which deliberately leaves them as bait for people who for two years, and even a small team, specialize in strange phenomena, consisting of two people, a physicist and an equipment specialist, and also ratified Tom's novel and work watch all night.

On the night of September 16, 1996, the team experienced the first vision of an airplane shape in the sky, moving in a strange way, but the images were not good enough, and after a year of observations, cow intestines appeared in minor terms, having lost the sound file without any rot or decay.

Continue to work to ensure that the team after this watch and they have their own work in earnest and they interviewed several people from this region, some gave the usual answers (curse of a demon or alien), and some refused to stop, in this case, investigate for eight years he saw in two men all sorts of visions, strange, but most of them were only for a moment and were not convincing to them and decided to end what he had achieved due to lack of evidence and ended Tom's story with this property.

The farm of "semantic" problems oscillates between the threads of mystery
Are you a haunted housekeeper or cursed ... a pending question.
Tom's story attracted the attention of many fans of the supernatural and the phenomena of a special and special airport, zip-plates of an airplane and believers in aliens, and the story that managed to spread quickly spread around theories of the supernatural:

- The Indian Curse Theory: And this is a theory that believes in the category of the Senate and the elderly in this area, they have always believed that the curse of melancholy hovers around this place.

- The theory of aliens: and the follow-up is considered safe because fear takes on the young fan of science fiction and special comics because they see Tom's lights in the sky.

Secret Experiment Theory: What is safe from the negative thinking of the US government and the experience of using a sophisticated new weapon that affects the brain of an individual, this family of innocents, which made them nervous, or that those lights in the sky are model aircraft designed for the military USA.

- The theory of civilization has been lost: the modern theory says that, perhaps, she lived in that region, the ancient civilization reached a high degree of development and what it is, if not its achievements or its remnants.

But no one thought that Tom wanted to draw attention and light to him, it was someone who was not known to lie or fabricate a story as he kept his family away from investigating what happened and from the media. information.

He claimed that many of the people who visited this place saw strange things, and UFOs, and halos in the sky, and this is so famous, but their stories were exaggerated and implausible.

I personally see it as an experience of the US Army, especially for the American desert known for its privacy rules and strange experiences.

It was one of the most unusual and most amenable to certification by the American society, which abounds in a strange way this phenomenon associated with UFOs and space creatures and turned the property of the Investigation Directorate of this phenomenon also had a film production, film, lifestyle and documentary photographer called skinwalker ranch and he has released many documentaries about him in the media.

And you, dear reader, what are you looking at? ..

The first black hole in history .. A door to solve one of the most difficult puzzles scientists have ever faced!

The universe is full of good things so strange that it is difficult for a person to understand them, especially since we believe that we can have complete control over the universe with the help of laws and physical processes, open new deals, make sure that it is impossible to completely control the universe or even interpret the phenomena existing in it and, over time, we begin to break the laws in vain and become unable to resist the extinction of the area and our understanding of the nature that surrounds us, we need time for new laws to be able to deal with space mines.

On April 10, 2019, the shooting of the first black hole in the history of mankind took place, this find is no less important than others, and is a very big step in the science of physics and in science in general. But there are still questions, curious, logical, some are perplexed about this, and you need to do this:

How to get a black hole millions of light years long? How to make the body invisible? And what is the use of the fact that just a photograph has no meaning for a black hole ??? ..

Today I will tell you, dear readers, about the newfound black hole.

The first black hole in history ... a door to solve one of the most difficult puzzles scientists have ever faced!
Black hole image .. it must be fully certified.
Initially, those who do not know what a black hole body is, do not see a very high density in space resulting from the explosion of a giant star in the process of a supernova, and begin the mass of the star product of the conjunction points in a very small space and have the ability to absorb stars, moons and planets and even light itself, given the attractiveness of a black hole, depends on infinity, therefore, things inside it are judged to stay forever, and it is impossible to lay a body to withstand this enormous gravity. Black holes that already exist and there is a lot of evidence to confirm their presence are being tracked step by step and astronomical observatories, and finally we managed to photograph, and you know about the presence of a giant black hole in the center of each galaxy, in addition to thousands of black holes. located at random points in this area.

All the photos that we see black holes, whether on the Internet or in films, articles or others ... it was just a forecast and simulation of scientists and there is no image of their real, but now you can get a picture of a black hole, it's in the millions light years involved in this feat of over 200 worlds and using 8 sets of telescopes found in locations on the earth's surface, although I still wonder why the scientific community was turned upside down just a picture was released and let me tell you that the scientific the community has been in the middle of the picture for over two years now. This is not just a picture, but also a key to one of the mysteries of the universe, and an answer to questions that have puzzled scientists for hundreds of years, and is considered a big step for humanity on its way to understanding the Universe ..

We will now answer a few questions. methodology..

How can scientists depict a black hole? How did they know about his existence from the very beginning, despite the fact that his body is invisible ?!

The first black hole in history ... a door to solve one of the most difficult puzzles scientists have ever faced!
Pictures taken by black hole scientists
Black holes, as we know, these bodies do not reflect light, and do not see something does not mean its absence, but enough that we can have with the help of other means besides the naked eye, this is the way that scientists with most theories and discoveries that emerged in the scientific community, starting bodybuilding was very small, such as corn, and at the end was very large, such as black holes. Corn electronics were golden, for example, bodies are invisible, but we know almost everything, because the conclusion about them, the inability to see something does not mean its absence, but it is enough to conclude that the same is exactly the case with black holes.

The first line in the photographic black hole project is that astronomers, they see a strange scene, observing them for some of the stars that are 26 thousand light years away. These groups were spotted by a team of researchers using infrared telescopes, and the scene looked like this:

These stars rotated in fixed paths around a specific point, and, oddly enough, if that point was a butterfly! These questions :

-Why do these stars revolve around the vacuum point?

-And the point of separation is already there, or is it something else that we do not see?

And it was the right time so that at this moment a vacuum did not form, but his body became invisible. it had a tremendous attraction, very capable of pulling stars around; tell me about Albert Einstein for over 100 years.

The first black hole in history ... a door to solve one of the most difficult puzzles scientists have ever faced!
I've been telling you about Albert Einstein for over 100 years
Astronomers have complete confidence in the existence of a black hole thanks to several clues, including the theories of Albert Einstein and the vacuum gas information I talked about earlier, and also because they track a huge amount of ultraviolet reflections of huge thermal radiation emitted from a point not visible. despite the disappearance of the black hole in the Universe, only now these things reveal its place. And since human nature is curious and his friend is only appropriating all aspects of the subject, the scientists decided to cut out the garbage and export the black hole, and therefore they subjected Albert Einstein's theory to the most severe test! If it turns out that the shape of the black hole is identical to the problems that I told you about Einstein, will the theory of relativity be correct, are there any mistakes, but if it turns out that it differs from the playoffs? All theories of relative error, which are considered a big catastrophe for the world of physics, will be sad.

The black hole that was shot from millions of light years away, and so we need a telescope, is huge, very, very much about photography, that's for sure, my friend. we want us to export flea waves to the lunar surface because in the black hole theme people are like ants who want to know what is on the other side of the river.

Entities can visualize a black hole, which is about millions of light-years away, we need a giant telescope very equal in size to planet Earth, almost, because these future scientists are scientific tricks: they filmed parts of the upper black hole through a telescope from different ones available around the floor, and then they compiled the images into one image in the end, it looks like a GA game when you put a car together in one place. Although the task is very difficult, but it showed wow, already starting in April, the research group began to work on the project, the idea was that the work done was the work of a network of telescopes around the globe, these youth were attached to eight groups and distributed over the entire planet, France and Greenland, as well as the United States, Spain and Mexico, Hawaii, Chile and Antarctica, so these young people telescoped an alternative giant telescope the size of planet Earth, which had already exported a black hole.

The first black hole in history ... a door to solve one of the most difficult puzzles scientists have ever faced!

Even light has no stones advantage of the sexiest hole
Another question that was asked by scientists - how can they photograph the body invisible?! ..

In fact, the black hole is invisible because its attraction is very strong and therefore absorbs light rather than reflects it, which is why it appears on the butterfly's body. The answer to this question is the presence of circular loops in the vicinity of the black hole, and these loops have a certain size and shape. There was a circular area inside this episode that is the hole itself, or be rings emitting around the hole, this is because objects and materials that enter the hole due to its terrible attractiveness, and this material enters too quickly, trying to enter them. collide with each other, which results in a very high temperature, almost twice the temperature of nuclear fusion at 14, which in turn causes beams of radio emission - rays in very large quantities, and this is the orange color around the black hole, which was tracked by radio telescopes.

In 1916, he predicted to Albert Einstein in his theory of general relativity the existence of bodies, the giant has a tremendous ability to swallow everything that is close to it, and considers their attraction to be the money end, and the name Black holes, predicted by the approved circular shape surrounded on the outside, has the ability to disassemble the body before it is swallowed by a black hole, resulting in a very formidable force, even if this body itself glows. Albert Einstein said that for over 100 years he has been doing without a telescope or any modern instruments, and instead uses pen and paper and some operations. Indeed, with the passage of time and the emergence of the modern era, clear evidence of the presence of these objects appeared, and finally we took them on film and found that this is the same design that he said about Albert Einstein and the laws of general relativity, which is a new victory Albert Einstein even after his death, it seems, Einstein's mind is one of the best minds in history, starting with ‏الThe GBS that's in your phone that's using Einstein's laws to find you now.

The images that the black hole displayed on the screen confirmed the validity of the theory of relativity, and this achievement is considered the door to solving one of the most difficult mysteries of the Universe and for scientists who are trying to connect the two theories that subject them to the motion of objects throughout the Universe.

This is the general theory of relativity, which regulates the movement of objects that are too large in space.

And quantum mechanics: which explained the movement of atoms and electrons was very small, and even now scientists are struggling to get to the link between the two theories and they are sure that black holes are the link.

Important note: This discovery is one of the most important discoveries in the history of mankind, after the announcements of news channels, as well as pages and sites on the Internet stated that black holes are located around the church, which is mentioned in the Holy Quran ayah in Sorat al-Islam (so I swear by diphtheria (15) around the church (16))

But from the point of view of my humble my dear brother, the Qur'an should not be associated with any scientific discoveries of recent times, because it remains theories and transactions of health and fitness; if we tie the Quran with the finds of a strong hook, we can show other theories or modifications in these discoveries, and perhaps some people think that the Quran is also wrong, the Quran does not need both theories and equations to prove your health.

- I wonder if you think that scientists will ever have a theory linking Albert Einstein's theory with the theory of quantum mechanics? Use what you can achieve using this theory? Of what?

-According to Einstein's theory that has been proven, do you think we can travel in time?

Tell us your opinion, dear reader ..

Andrew Cross: how can an organism be born from stones?

For more than 180 years he ruled the world, which he calls (St. Andrew's Cross) (creating) life, as through one of his experiences that took place in his castle, well, it is not called that, and he did not say that he did not do anything contrary to the teachings of the Church, but he could not ultimately explain where these beings came from, which appeared as a result of his experience.

But who is the St. Andrew's Cross? And what is the history of his conflicting experience, and what are the creatures who found it? Come to find out the whole story.

Andrew Cross: how does an organism that is born from rocks work?
Andrew Cross was born in the 17th month of June 1784, his father died in 1800, and then his mother also died, also in 1805, he inherited the Andreevsky house of a huge family and its vast territory. The analysis of the music room in the Palace his big to the laboratory for the patenting of numerous and accepted by people at that time madmen came from a mad man, and he closed his experiments with electricity, and sparks of electricity flashed in his house, and you can imagine, dear reader, a home made from a light source (anonymous) for people at the time, what they think of ordinary people, still think lanterns in lighting, this is the price of right.
It was Andrew's cross, who is reliving in the hope of inventing a new metal, but those experiences took on an even stranger curve and made something meaningless.

In his book entitled: (Memorials, Scientific, Literary or Own St. Andrew's Cross), published in 1857 after the death of a person a few years later. I wrote to his wife (Cornelia) the following:

"In 1837, when he was Mr. Cross, he indulged in one of his own experiments on the formation of electricity, and during these experiments attracted the attention of insects who found themselves in circumstances fatal to living things. And did Mr. Cross not only write down that found: any real world, a renewed religion, regardless of its unscientific nature and discoveries and codification of everything small and large with high precision. Despite the fact that the appearance of these creatures was unexpected for him, and if it was impossible to explain the reason for their appearance, neither one theory, however, to interpret the appearance of a stranger in this way. "

Andrew Cross: what about an organism that is born from rocks?
The house in which he lived faced experiments throughout his life

What is this trade that Andrew cross did and what led to the appearance of such creatures or insects?

The experiment was a mixture of water with potassium silicate and hydrochloric acid, which was rasterized on volcanic rock, which was constantly exposed to electric current through two wires tied to a battery with a voltage in a chemical solution. And the days passed without any significant result, and the cross began to feel despair and disappointment, when I noticed something unexpected in the liquid that the trade carried, instead of the expected crystals appearing, I found what looks like little bugs or fleas in Mobile.

Andrew Cross: how does an organism that is born from rocks work?
Image of an alien insect that appeared - At first I thought that the rock, or maybe the liquid used in the experiment, were insect eggs, microscopic and electrical stimulants to make them hatch. But the stranger in this item never hinted at or followed any eggs or found any egg rind after the insect emerged, and when checking containers containing liquid chemicals used in the experiment, found no trace of insect eggs or the insect itself. And it enables the curious world, which he spoke about when returning the trade, to reuse closed containers to prevent any possible contamination from the outside, and got the same results: the appearance of insects clearly occurs in the containers used in the experiment.

Books about her intertwined: "I have a complete insect, the reality of those on her tail, and she had other insects, within two days after observing these insects and they began to move their legs, and during the weeks of fasting, hundreds of these insects, and they began to crawl on the table and hide in places where they might hide. "

When he repeated the experiment, using several liquids that were toxic and known to be impossible to stop life in them, in most experiments this insect reappeared.

It looks like there is life out there, perhaps originating from rocks and chemicals and sent from a weak web !!

It was a loss of order crossover for communicating with the London Electricity Association The beetle gives his answer did not come into his experiments and informed his two best friends about American trade ...

Reaction to trade

Andrew Cross: how does an organism that is born from rocks work?
Duty to trade rejected by the scientific community and
Unfortunately, I have not found any answer to the cross from everyone other than the collapse of this, the story leaks out about his friends in the desert and I see what these insects stand for (Acarus Crossii lacrosse Croce) and I feel that the public and the people they began to call me an atheist, and he is a demon, and he is not just a person trying to help the same Creator God (create) the life of nothingness. And people began to go to his

palace, sneer at his work, and to destroy his reputation, and work to drive out the devil in front of his house, but they threatened him with force.

But he defended himself by the fact that he himself does not know much about how, as a result of his experiments, this happened, and he did not make any statements about the creation of life out of nowhere, he simply wrote down his experience, as he received, trying to get to the bottom of the reason for their appearance, and still doesn't know where these insects came from ... and all I wanted to achieve was to get to the crystal formations, but instead to the back of that crystal (insects) ....

Other scientists repeated the experience of the cross, and they got similar experiences, including Mir (and writes) made extra efforts to make sure that the experiment took place in an isolated environment and obtained the same results, but he and the other scientists who conducted the experiment were not targeted for the same protection as a crossover against defamation.

Andrew Cross: what about an organism that is born from rocks?
Keeps what happened during his experiments puzzle gas scientific research
Additional fire attacks on him and his experiences from religious people, stated by others that the cross bully eggs are microscopic deliberately in the liquid used in the experiment to mislead public opinion.

This led to pressure from the people and society, religious and scientific, who became a cross-prisoner in his house and speak to him openly, because he shows respect for people every time. Matt Cross in 1855 and didn't know what he found. or what it was, and did not understand why people accuse you, how terrible it is.

Trade has become a forgotten and forgotten world of St. Andrew's Cross and its experience for more than 180 years.

The question remains, who often visits hurt two, what did he discover (St. Andrew's cross)? The strangest thing is that he does not have any re-experience in our modern era, and that is why no one knows what will be in those containers in the laboratory at the English Court in more than 180 years.

Note: the story that I read in the book of his name (our world of the unknown), and I cannot find a mention of him in any other book, speaks of mystical things, and after a very long search, I managed to find the story of the original, and there is a page on Wikipedia about it.

The girl who slept for 32 years and puzzled scientists

In my country, Don, in southeastern Sweden, in a remote isolated town, born in Carolina, Olsen I read October, the girl was natural and beautiful, had to reach the age of 14, and put up on February 18, 1876 ..

32-year-old dormant girl confuses scientists
She lived in a Swedish town.
On that day, the carol came from home to spend time in the picturesque nature and in the fresh air, and during the crossing on the ice across the frozen river, she fell and almost drowned after she was hit with her head on the stones of the river, but was dragged to a safe place in the last moment, it was a small wound that quickly recovered from it.

A few days later, I suddenly began to complain of severe pain in the tooth, kept the pain with me for most of the day, so much so that her family thought that the inflammation and pain in the tooth were caused by magic, the inhabitants of the town on the then earth believe in witchcraft and superstition in to a large extent, and that any damage can be due to these things.

On the night of February 22, OTT went to bed with her, and she still has a toothache; us is a girl for a long time, longer than the usual bedtime, and when her family got worried, no matter how it was Maya's star, which could affect our health, they went to wake her, but she could not.

Hours passed .. days .. and Carolina is on the same course, do not wake up with the fact that she is alive and can breathe normally. I tried to wake her family up in various ways until they came

Use the pins to poke her in the legs, but the girl remains in a deep sleep.

Meanwhile, her mother was always there, replaced by fear and anxiety for her daughter, and discarding her decisions and domestic violence, Carolina slept and barely opened her eyes no matter what.

A 32-year-old scientist hibernating before is confused
Otto's bed never woke up like the Sleeping Beauty princess!
He was the father of Caroline, a fisherman, a poor man who could not afford to hire a doctor to turn to his daughter, the first in the family, and not on the advice of neighbors, and advice is the country ..

But all these advice did not lead to anything, after more than ten days neighbors came in between them to collect the doctor's salary, and the doctor was indeed called and went to visit Carolina for a whole year, and then wrote to the editorial office of one of the leading medical journals of the States of Scandinavia, so that in due time he helped doctors find a cure for this girl. After that, he visited several doctors, they noted with the girl that everything was going on as usual, namely that her hair and nails on her hands and feet had stopped growing, despite the fact that time had passed.

In the end, could these doctors diagnose the girl's condition - all attempts to wake her up failed, Caroline's family said that it was on rare occasions, you sit down and mutter indistinct words, and then go back to bed.

One of the kids, who visited the girl in such a state as she had some kind of hysteria, was transferred to a hospital for mental disorders, they used electric lightning, but this method also failed, a month after she was discharged from the hospital without waking up in her case, this time the doctors said that her condition could be a form of dementia, paralysis, despite the fact that there were no signs indicating that they were suffering from symptoms of this disease.

Finally, the doctors' despair in the diagnosis of the girl's condition was precisely related to the fact that Carol was suffering from a mysterious disease that they could not identify, and all this time Carol was in a deep sleep, her sponsoring mother gave two glasses of milk a day and some money in local currency and sometimes imprisoned that his mother died in 1904 around the world. After her mother died, the sponsored lady of the country volunteered to stay with her.

Two years after her mother's death, her brother died in 1907, the strange thing is that after these deaths, Caroline's intervention in a fit of tears where she cried when her mother died, also went into a fit of hysterical crying when her brother died, but she didn't wake up.

A 32-year-old hibernation girl confused by scientists
A rare photo of Caroline after she wakes up.
After many years of her long sleep, I woke up in Carolina's world rather changeable and in 1908, after she slept in 1876, Carolina was slim and pale too and had difficulty in controlling herself, showed sensitivity to light, could not identify the remains of family members , was shocked when she learned about the death of her mother and brother. Although Caroline had a rough time in college, she said she still remembers reading and writing.

The news of the awakening of all of our people inside the country spread like wildfire, and all people began to go to see her, because they know her history, as press correspondents from all over Europe, the USA and the USA came to countries to see her and become this story is very famous in the world.

When asked what she can remember, she replied that she remembers everything before falling asleep, she observed strange visions in the presence of blue immersed in water and exotic for her brothers who died by drowning.

And I took Carolina to check the mental and psychological in Stockholm, and everything was fine, where she did not suffer from any mental illness - this is health, mental and physical good, a miracle for Carolina, when I woke up, she was 46 years old, but she looks like a girl in her twenties, as if time passed more slowly during their long sleep.

A 32-year-old hibernation girl confused by scientists

The image of Caroline at the end of her life. After a while she wakes up. I started Caroline and her family feel uneasy about the intensity of the arrivals to visit the area, so I decided to lay low and live in peace for the rest of my life.
In fact, a large number of doctors did not believe Carolina's story, and they said that they probably agreed with her mother in making this story, and that she suffered from psychological problems and pretended throughout this period when this was their opinion about science and not everyone can save life for so long, and he feeds only on milk and an aqueous solution for work.

But if we assume that Carolina agreed with her mother, would agree with the woman who looked after her after the death of her mother, then this woman, I swear to you, in the past four years has not seen Caroline awake or repeated, except for incantations that you spoke vaguely after the death of her mother and brother.

In 1912, The meet the professor and psychiatrist Harald press Carolina published a study and analysis of their condition entitled: "sleeping on .. 32 years of fear", where he confirmed that she had suffered from deep sleep for 32 years for some for a mysterious reason she hadn't woken up for so long and that she was infected with a very mysterious keeping her eyes closed, but she interacted with grief and anger during her sleep, and this was no trick.

After that, Carolina lived a happy life, fucked hard until she died in April 1950, at the age of 88.

In fact, this strange situation remained scientifically undiagnosed with certainty and it was not known what was the reason that Carolina had been inactive for 32 years, despite all attempts to even be given an electric shock.

If Carolina is truly a miracle of medicine, then it is believed that this phenomenon is the longest sleep period for a person who lived on Earth.